NATURE AS TEACHER

NATURE AS TEACHER

How I Discovered New Principles in the Working of Nature

by
Viktor Schauberger

translated & edited by
Callum Coats

Gateway

Gill Books
Hume Avenue, Park West
Dublin 12
www.gillbooks.ie

Gill Books is an imprint of M.H. Gill & Co.

© The Heirs of Viktor and Walter Schauberger
© Introduction and Notes by Callum Coats

ISBN-13: 978 18586 0056 7
ISBN-10: 1 85860 056 1

Set in 10.5pt on 12.5pt Palatino by
Character Graphics (Taunton) Ltd
Cover design by Synergie, Bristol
Printed and bound by Replika Press Pvt. Ltd., India

The paper used in this book is made from the wood pulp of
sustainably managed forests.

A catalogue record is available for this book from the British Library.

15 14 13 12

Contents

Introduction

At the time of writing the world is being engulfed by increasingly cataclysmic manifestations of the disturbance and disruption of Nature's otherwise orderly processes. From reports received almost daily, both nationally and from around the world, we are increasingly forced to become aware of certain life-threatening irregularities in the functioning of Nature's household. Record catastrophes of increasing violence and extent are being reported in almost every country; tornadoes, deluges, widespread flooding, searing drought, earthquakes, unseasonable snowfalls and extremes of temperature, all of which are associated with huge loss and suffering. Strife and starvation are on the increase, coupled with a seemingly endless emergence of hitherto unheard of diseases. The majority of these existence-threatening events, these so-called 'natural disasters', are not of Nature's making. On the contrary, they are directly attributable to the misdemeanours of humanity, the result of its arrogant repudiation or even total ignorance of Nature's sublime laws and the subtle interactions between the all-permeating interdependencies upon which all life is founded.

To this catalogue of climatic irregularity must be added the more directly apparent man-made factors. The world's river systems and oceans are gradually collapsing through pollution with chemicals. Fish and other aquatic life are dying. Other creatures are being threatened with extinction or have already become extinct. The environmental overload is being increased at a precipitous rate through excessive land clearing, uncontrollable forest fires, the reduction in the quality of light reaching the Earth's surface due to atmospheric pollution, and the saturation of all living things, down to the smallest cellular organisms, by the cocktail of electromagnetic emissions known as electrosmog. This undoubtedly has a disturbing effect on the bioelectric and biomagnetic information that controls the proper functioning of the cells' delicate metabolism, which in aggregate leads to physical disorder and abnormality. Not only does this affect our physical well-being, but also our behaviour and mental abilities, thus inaugurating a decline in morals and the capacity to think creatively. According to Viktor Schauberger a brain, whose physical constitution and intellectual power has thus been corrupted, would be incapable of comprehending Nature's causal dynamic interdependencies.

As a result of humanity's misguided activity, the Earth is becoming increasingly unstable; an instability that is also reflected in the growing instability of human institutions. These alarming events, however, will continue to increase in scope and magnitude in a manner presently inconceivable, unless effective remedial measures are instituted on a large scale as soon as possible.

Since Nature's processes are interdependent, the total synergetic effect of the present disruptive activities can reach a magnitude out of all apparent proportion to the strength of the individual contributive stimuli. If the Earth is viewed as a sphere with a diameter of 1 metre (1,000 millimetres), for example, then the mesosphere, which largely defines our living space, would extend to a height of about 10mm. As a living organism, the Earth breathes and pulsates. In common with other living creatures, and symptomatic of a sick or diseased condition, physical convulsions cannot be ruled out. Therefore, viewed at this scale, even a barely perceptible ripple 1 mm high swirling around the globe is not inconceivable. In real terms, however, this would signify a tidal wave about 6.5km (4 miles) high. Such a tsunami would mean oblivion.

It is necessary to write this down to emphasise the precarious position in which we find ourselves today, so as to provoke a greater interest in rectifying matters urgently. If we wish to reverse this downward course, and to do so sustainably, then we must learn to feel Nature once more; to harken to her voice and to take heed of her subtle movements. Some of this natural awareness, now long forgotten, we must learn from the very beginning. We need, as Viktor says *"to perceive the world not as an action, but as a reaction"*[1], for only thus will we be able to arrive at the true causes of things and the reasons for the present parlous state of the world. While some causes are slowly being discovered, the underlying reasons, which are the most important and have the greatest influence on outcomes, have so far not been addressed.

But how have we arrived at this precarious situation? In the 16th century, before the development of science and technology, people were generally more attuned to the cycles, pulsations and subtler movements of Nature's energies. With few technical aids and artefacts, it was a more organic society and more intimately embedded in the natural world than modern humanity. People had a far more sensitive awareness for the events and movements of natural energies. With the gradual rise of science, this close connection with natural phenomena and their perceived causes was increasingly lost as the materialistically based methodology of science grew in scope and application. Science only interested itself in what it deemed to be the physical

---

[1] From "Return to Culture" in the Schauberger archives. – Ed.

causes of a phenomenon, in what could be observed directly. It attributed causes to events which were actually the effects of causes unseen. As the sophistication of science increased, so too was its widening estrangement from Nature and lack of real appreciation of her underlying patterns. In pursuit of this rational, materialistic approach, a language and technical terminology gradually evolved, which, while able to describe with great precision the mechanical and physical effects being studied, was incapable of describing the true inner conformities of natural law. Science was thus able to describe the outer effects, but not the real inner causes of natural phenomena. One of the principal victims of this inability is the lack of understanding of the essential nature of water – as the Giver of Life – and how it should be treated. Through all the disasters mentioned earlier, however, public attention is being drawn more and more towards the importance of water, its availability and above all its quality.

The purely economic paradigm, the offspring of a materialistic science, that governs all humanity's present activities is untenable in the long term, for no economy of whatever kind is sustainable unless there is an abundance of clear, clean water and thriving vegetation. This is, and has always been, the bottom line of existence and evolutionary development. The thrust of the economic paradigm, while expanding and making communication more instantaneous, which has made people aware of global events as they happen, has its downside in the rising fascination and belief in the power of technology.

The increasing pursuit of 'hard-edged' technology and science led to our increasing obliviousness of natural energies and processes. Therefore our language never developed expressions for natural things intuited, sensed or otherwise experienced by those more naturally attuned. Despite our much vaunted, yet one-sided technical sophistication, we have in effect become more primitive than the so-called 'primitive' peoples who live in close contact and harmony with Nature in primeval forests. This inadequacy of language presented Viktor Schauberger with considerable problems in accurately describing all he perceived, for as he said *".....it is exceedingly difficult to differentiate contemporary concepts from those required here, for which there are no technical terms. ....Since no exact terminology exists, complicated paraphrasing is required to describe the higher origins and causes of motion and formation. ....Therefore modern (and frequently inappropriate) terminology will have to be used. By placing them between quotation marks a different sense or meaning is intended. This I can do nothing about. Later on, other and fresh words will have to be coined to describe these new concepts."*[2]

Despite his attempts at the greatest possible clarity in his explanations, he was the first to admit, however, that *"Few will understand the meaning of what*

---

[2] From "Organic Syntheses", *Implosion* Magazine, No.22. – Ed

I have stated! Some individuals, however, will obtain an indefinable inkling..[3]*....Only those with exceptional intuitive abilities, and thus of an artistic turn of mind, will actually be able to grapple with this extremely difficult way of thinking.*"[4] Some of these new words have been coined and where they occur in the text their meaning is explained.

There is now an increasing urgency to rehabilitate this planet while there is still time. For this, a new and far profounder knowledge of Nature is necessary, so that whatever is implemented by way of remedial measures, will be in harmonious accordance with Nature's laws. Fortunately, a large part of this fascinating new path has already been trodden and in the following pages we will be able to accompany Viktor Schauberger on his voyage of discovery into realms of energy and natural phenomena that hitherto have escaped our senses. The greater the number of people who become aware of these new possibilities and actively pursue them, the greater will be the force to bring about their urgent practical implementation, now more than at any other time so desperately needed.

And so in the pages that follow we will obtain entirely new insights into Nature's mysterious inner workings. These will open up a plethora of new possibilities for understanding how to reorganise our activities, so that they harmonise with natural law. We will see how it was that Viktor Schauberger seems to have been guided unerringly to being at the right place and the right time to see otherwise unimaginable events unfold. In witnessing them, he gradually became able to perceive and understand the higher formative and controlling energies that organise and dynamise physical existence. In this way he was able to put together, piece by piece, an entirely new environmental paradigm, which will do much to help extricate ourselves from an otherwise irreversible decline.

It was the gradual revelation of these new concepts as I was translating them that continually motivated me to translate as much material as I could lay my hands on. I felt that all was not irretrievably lost; that there was a way out of the global predicament. Here at last, there seemed to be answers to many problems that confronted human existence. The desire to discover more, constantly fired my enthusiasm. I developed a thirst for this new knowledge which I sincerely hope will also galvanise the reader as much as it did me.

Callum Coats, May 1998.

---

[3] From the Special Edition of *Mensch und Technik*, vol.3, section 7.4, 1993. – Ed.
[4] From "The Trout Motor", *Implosion* Magazine, No.111. – Ed.

Sources

The sources of the articles in this volume are as follows:

Our Senseless Toil
Written by Viktor Schauberger between 1932 and 1933, it was originally published in a two-part book entitled, *Our Senseless Toil – the Cause of the World Crisis*, subtitled *Growth through Transformation not Destruction of the Atom (Unsere sinnlose Arbeit – die Quelle der Weltkrise. Der Aufbau durch Atomverwandlung, nicht Atomzertrümmerung)*. Part I first appeared in 1933 and Part II in 1934. Both Parts I & II of *Our Senseless Toil* were published by Krystall-Verlag GmbH., which due to financial difficulties was finally closed down in April 1939 by its then editor-director Franz Juraschek.

Mensch und Technik – naturgemaß
Originally *Kosmische Evolution or Cosmic Evolution*, the German periodical, *Mensch und Technik – naturgemaß (Humanity & Technology – in accordance with Nature)*, is funded by private subscription and published by the *Gruppe der Neuen*, (the New Group), whose aim was to explore Viktor Schauberger's theories and to interpret them scientifically. Volume 2, 1993, is devoted entirely to the recently discovered (early 1990's) transcript of a notebook compiled in 1941 by a Swiss, Arnold Hohl, which reports on his visits to Viktor Schauberger in 1936 and 1937. It is from this volume that the passages in this book are obtained.

Implosion
Implosion is a quarterly magazine, funded by private subscription and generally oriented towards the lay reader. It was originally published by Aloys Kokaly from about 1958 and now runs to 122 issues. Kokaly also founded the *Verein zur Förderung der Biotechnik e.V.* (Association for the Advancement of Biotechnology) specifically for the research and evaluation of Viktor's theories and through *Implosion* to provide a platform for Viktor Schauberger's various writings, of which Kokaly had many originals.

The Schauberger Archives
Forming the greater part of Viktor Schauberger's estate, these are the private archives of the Schauberger family and the PKS (Pythagoras-Kepler-School) at Lauffen, near Bad Ischl in Upper Austria.

Our Senseless Toil (1933)[5]

"Viktor Schauberger often commented on his ideal picture of free people which he took from the more evolved wild animals who, in a free and undisturbed Nature, know no hunger or dependency. 'No stag greases another's palm' – that is, there is no dependency and no exploitation."

(Aloys Kokaly, *Implosion* Magazine, No.47, p.14)

The Achievements of the Twentieth Century

We have today become accustomed to the fact that millions of people can no longer earn their daily bread by honest labour. They must obtain the necessities of life by scavenging from refuse heaps like animals, by begging, robbery, fraud or even murder. Having lost faith in current methods and customs, and viewing education as an exercise in futility, our children are banding together, arming themselves and preparing to secure their rightful place in society by force.

Under conditions where hospitals and refuges for the homeless are overcrowded, where the clientele of drug addiction clinics and lunatic asylums is increasing and where cases of suicide are growing; it is understandable that no sane-minded person any longer gives credence to the empty promises of our leaders. It has long been apparent to thinking people that, if no radical departure in current economic doctrines and practices occurs, then only an enforced decimation of the overlarge population, a well-organised mass-murder or, expressed more aesthetically, a modern war, might perhaps cut this Gordian Knot.

The most remarkable aspect of the matter is that neither the instigators of this apparently inevitable war (who seem concerned only for their own skins) nor our young people (who seem determined to fling away what they perceive as

---

[5] "Our Senseless Toil" is the longest of Viktor Schauberger's published writings. It is not reproduced here in full, however, as some portions have been relocated according to theme in other volumes. Those relating more specifically to water in *The Water Wizard* and those to forestry in *The Fertile Earth*, respectively volumes 1 and 3 of the Ecotechnology series. Some short items from elsewhere, whose sources are indicated, have been included in this introductory section, which finishes with the conclusion of "Our Senseless Toi". – Ed.

their futureless, worthless lives) have recognised that this sacrifice is totally uncalled for. In other words, the ghastly, almost suicidal annihilation of the despairing masses with poison gas or other weapons is absolutely unnecessary.

Both parties are unable to perceive that this scenario, which is impossible to equate with any cultural development, is only the entirely natural secondary effect of the intervention of a much higher power; a power with completely different means at its disposal. Means that are more fundamental, and more importantly, work much more drastically than all the weapons of war human brains have ever devised. If we were being honest, these were the very brains that we all had hoped would find a way out of this chaos.

Let members of the older generation philosophise, drawing attention to their reductionist knowledge and its hollow clichés. Let them rant and rave over these developments, for in the final analysis, the rising generation is absolutely right no longer to place any trust in the ability of its forebears, who have borne such bitter fruits. They are right to refuse to follow the false dictates of our intellectual leaders, who have brought such wretchedness upon us. Despite its supposedly high technological culture, the civilised world has reached such an ethical nadir that it has become incapable of perceiving that this physical and moral decline is none other than a progressive cultural dissolution.

Humanity's most sacred possession, its freedom of subjective thought, action and feeling, will be literally trodden underfoot by people who were never really in a position to intervene in a positive way. In such a situation the colours under which these leaders choose to march is quite immaterial, because the same oppressive drive exists everywhere. Generally, the inner perception of the true causes has been lost, and consequently, the last chance of really effective help. Moreover, those in positions of power, who are incapable of forming their own opinion, must constantly rely on the advice of so-called *experts* who are themselves victims of a universally inferior education. As a result, they are unable to realise that it is precisely their advice, and the actions arising from it, which will inevitably transform this Earth into a hell, when it could be a paradise.

If humanity does not soon come to its senses, and realise that it has been misled and misinformed by its intellectual leaders, the prevailing laws of Nature (with poetic justice) will reliably act to bring about a fitting end to this ineptly contrived culture. Unfortunately the most frightful catastrophes or scandalous disclosures will have to happen before people realise that it is their own mistakes that have led to their undoing. These can only be rectified with great difficulty precisely because they were principally committed by the authorities. Rather than pass judgement upon themselves, these institutions and individuals, who are ever protective of their own interests, would allow millions of their fellow human beings to perish before they would ever admit to their mistakes.

In discovering the causes, however, only one small step has been taken, because a host of so-called experts is arrayed against any systematic attempt to put these errors right. These experts are obliged to advocate the course they have championed, because it is their livelihood and they wish to be looked after until the end of their days. Yet, even this obstacle might be overcome, if the mistakes could at least be restricted to a particular branch of industry. A thorough investigation into the most common mistakes made over the centuries reveals the enormous spread of the malaise arising from fallacious precepts and perverse practices. It reveals such grave cultural, technological and economic transgressions that no branch of industry is left untouched. Not even a partially unaware expert, can absolve himself of his complicity, whatever his chosen field.

At the outset a powerful opposition must be reckoned with. It will be quite futile to expect any support from experts when, under these circumstances, it becomes evident that nearly every one of them would be threatened. But this obstacle should cause no alarm, for we are not concerned here with the livelihood of a few, but with the existence or non-existence of the whole of hoodwinked humanity. The behaviour of our young people today certainly provides clear evidence that humanity is still morally healthy. They militate vehemently against the signs of decay emerging everywhere and refuse to continue to trot mindlessly down the road to war that has led us into an economic and cultural cul-de-sac.

Opposition alone, however, achieves nothing. Our youth will only achieve any practical success in their struggle when the causes are identified and the errors are revealed that we and previous generations made, so plunging the world into misfortune. For this reason it will become a sacred duty for all those who perceive the full extent of what has happened, to put aside all personal advantage and enlist for the final putting-to-rights of these many errors. The same is also the duty of everyone whose inner feelings admit the mere possibility of wrong-doing.

The most effective way of righting these wrongs is to inform the general public of the great dangers of defective reasoning, and the futility of pursuing the present goals. Rich and poor, high and low alike, must become seized by doubt and well-founded mistrust. Affecting ever widening circles, this will ultimately kindle an inner sense of self-preservation in the broad mass of people. Once awakened, this inner sense must not be allowed to rest until the people (and therefore God) have made their verdict known. They will then begin to work at a grass-roots level and bring about the necessary change for the better. It may indeed be a thankless task to inform the broad mass of people of the coming dangers which it neither sees nor wishes to see. However, regardless of the possible futility of revealing the fearful *Menetekel*[6] hovering above them, the

---

[6] 'Menetekel': Doomsday vision or portent of doom. – Ed.

attempt should still be made. At least our children and those dying helpless-ly in hospital should be made aware that they are the victims of history and the present culture arising from it.

Therefore the purpose of the following discourse is to arouse this inner sense of self-preservation in the public. If an instinctive premonition of the enormous dangers ahead, coupled with the mistrust latent in every human being, can be successfully awakened; then neither the problems affecting the Establishment's prestige nor its fears for its future will significantly impede humanity's final rescue from self-destruction. It is not the purpose of these general explanations to elaborate on the many indicators brought to light in a review spanning a thousand years. These are referred to briefly only where they have a profound significance and their correct understanding is necessary for comprehension of the whole. Naturally with such understand-ing, much will also have to be discarded. Once humanity perceives the won-derful conformity in natural law, and the uniformity prevailing throughout Nature, it will gain ethically and renounce any over-reliance on outward appearances.

If we want to influence the course of our own existence positively, an exis-tence now constantly imperilled by the re-emergence of alien life-forms, and if we wish to safeguard it against further degeneration, we must allow Nature to take command. Or, if we do wish to intervene, we must first become conversant with the simplest principles of Life. Every living thing is ultimately a bridge towards the build-up of the whole. Similarly the various religions and world-views merely represent spiritual bridges (often in prim-itive form) and so must make way for better ones, once the ethical upswing of humanity has overtaken them. Indisputably the mightiest bridge of all for the evolution of life is represented by the entity *WATER*.

Science views the blood-building and character-influencing *ur*-organism[7],

---

[7] '*Ur*-organism': In Viktor Schauberger's writings in German, the prefix '*Ur*' is often separated from the rest of the word by a hyphen, e.g. '*Ur*-sache' in lieu of '*Ur*sache', when normally it would be joined. By this he intends to place a particular emphasis on the prefix, thus endowing it with a more profound meaning than the merely superficial. This prefix belongs not only to the German language, but in former times also to the English, a usage which has now lapsed. According to the Oxford English Dictionary, '*ur*' denotes 'primitive', 'original', 'earliest', giving such examples as '*ur*-Shakespeare' or '*ur*-origin'. This begins to get to the root of Viktor's use of it and the deeper significance he placed upon it. If one expands upon the interpre-tation given in the Oxford English Dictionary, then the concepts of 'primordial', 'primeval', 'primal', 'fundamental', 'elementary', 'of first principle', come to mind, which further encompass such meanings as: – pertaining to the first age of the world, or of anything ancient; – pertaining to or existing from the earliest beginnings;- constituting the earliest beginning or starting point;- from which something else is derived, developed or depends;- applying to parts or structures in their ear-liest or rudimentary stage; – the first or earliest formed in the course of growth. To this can be added the concept of an '*ur*-condition' or '*ur*-state' of extremely high potential or potency, a latent evolu-tionary ripeness, which given the correct impulse can unloose all of Nature's innate creative forces. In the English text, therefore, the prefix '*ur*' will also be used wherever it occurs in the original German and the reader is asked to bear the above in mind when reading what follows. – Ed.

water, as a chemical compound, and supplies millions of people with a liquid prepared from this standpoint, which is everything but healthy water. All efforts to make science acknowledge the serious errors it has made are useless from the start, because for it to make such an admission would surely be to condemn itself. Of necessity, therefore, it must adhere to its present doctrines. All those still possessed of healthy common-sense should categorically refuse to continue to drink water prepared in this way. By continuing to consume such water they will inevitably degenerate into cancer-prone, mentally and physically decrepit, physically and morally inferior individuals. Let the experts and scientists, who are heavily attacked in what follows here, examine everything objectively and refute, if they will, the many criticisms presented.

Those best placed to judge whether these assertions are well-intentioned or not, are the farmers, already struggling so hard for their native soil. Let all those who are forced to work in the great cities seriously contemplate what would happen if, in addition to their bread becoming increasingly scarce, expensive and of worsening quality, their water also disappeared.

This danger will be all the more dreadful because the remaining reserves of water will become an unquenchable source of that most frightful disease, cancer. Cancer is constantly on the increase and, if too far advanced, has no really effective cure. Therefore let all those not fortunate enough to enjoy a cooling drink directly from a healthy spring consider where their water comes from, how it is distributed, and what artificial additives are used to make it drinkable.

Those unfortunates who are forced year-in and year-out to drink sterilised water should earnestly consider how an organism will be affected by water whose naturally-ordained ability to create life has been forcibly removed by chemical compounds. Sterilised and physically-destroyed water not only brings about physical decay, but also gives rise to mental deterioration and hence to the systematic degeneration of humanity and other life-forms. The same is equally applicable to all forms of vegetation and all other preconditions for life in Nature. The reason people mistake their cultural and economic decline for a passing crisis and strive in vain to master the increasingly widespread misery, lies mainly in the intellectual deterioration of humanity. Conforming to natural law this deterioration is followed or preceded by physical degeneration. Only a penetrating study by intuitively gifted people can fathom the innermost nature of the life-giving substance, *water*. Only through a painstaking investigation of the materialised ur-substance, water, will it become possible to show a mentally and physically degenerating humanity the ways which will once more lead us upwards.

Progress through Transformation of the Atom – Not its Destruction!

By means of only slight variations in temperature I have succeeded in decomposing various substances (elements and their compounds, minerals, metals, *etc.*) into their constituent parts, and subsequently to rearrange and recombine them. At present the scope for practical application of this discovery cannot be assessed, but it would undoubtedly imply a total reorientation in all areas of science and technology. Using this newly-discovered conformity with natural law I have already constructed fairly large installations in the fields of log-rafting and river regulation. They have functioned faultlessly for a decade and today still present insoluble enigmas to the various scientific disciplines concerned. Present management systems of forestry, agriculture, water and energy resources, as well as many theories and tenets of physics, chemistry, botany and geology will have to undergo a radical departure from basic principles. Even medical science will not be left unscathed by this discovery.

In this way it is possible to generate any amount of energy in and from water itself and to regulate watercourses over any given distance without embankment works. It is possible to transport timber and other materials down the central axis of flow, even if these materials (ore, stones, etc.) are heavier than water. It is possible to raise the height of the water table over a whole region and to endow the groundwater with the full spectrum of elements required for the prevailing vegetation.

Furthermore, timber and other materials can be rendered incombustible and rot-resistant. Drinking water and spa-water of any desired composition and therapeutic effect can be artificially produced for man, beast and soil, in the same way that this occurs naturally. Water can be raised vertically in pipes without pumps. Electricity and radiant energies of any magnitude can be generated almost without cost. Soil quality can be improved and cancer, tuberculosis and nervous disorders healed.

The Disrupted Cycle – The Cause of the Crisis

Today a yearning for living naturally is on the increase. This craving for a strong, peaceful and healthy Nature is an inevitable symptom of the present age and the counterbalance to the inorganic civilisation we erroneously describe as *culture*. This civilisation is the creation of humanity, who high-handedly and without consideration for the true workings of Nature, has created a world devoid of meaning and foundation. Now Nature threatens to destroy humanity, for through his behaviour and his activities he, who should be her master, has disturbed Nature's inherent unity.

Today we are standing helpless and perplexed before all that we have created; increasingly forced to recognise that our work, with all its problems, merely serves our own self-destruction. With no glimmer of improvement anywhere in sight we feel hopelessly propelled towards a forlorn future. It is quite understandable therefore that an increasing number of people are sick and tired of this insane activity and now seeking ways to return to Mother Nature.

The human is a being created according to Nature's laws and is therefore dependent upon them. In the course of time our *magnum opus*, our self-created pseudo-culture,[8] has become a meaningless and incoherent monstrosity. Through the immense power of technology it has reached such gargantuan proportions that it almost equals the power of Nature herself. At the very least it is already able to interfere destructively with her great life-giving functions. Humanity represents but a small spark, a mere microorganism in Nature's great panoply of Life. Encouraged by a short-lived, illusory success, humanity has embarked on a course that is beginning to disrupt the great coherence of Life. Not only this, but it is also about to put an end to all high-quality growth and production on our macro-organism, Earth.

Despite our accumulation of material wealth, humanity is now engulfed by a widespread economic collapse. Many areas of production exhibit regressive trends so that visible epicentres of decay are increasing on all sides and threatening humanity itself. Despite all the research no means can be found to prevent humanity from decaying alive. This is no more than the just and legitimate consequence of human activity. Knowing nothing of Nature's omnipotent laws, and with mindless greed, humanity claws into the life-giving organism of Mother Earth. She is now, with elemental power, beginning to paralyse the wanton hand that dared disturb the forces that serve all Creation.

This unique law, which reigns supreme throughout Nature's vastness and oneness, expresses itself in every creature and organism. It is the Law of Ceaseless Cycles that in every organism is linked to a definite time-span and a particular tempo.

If some intervening force should either accelerate, retard or altogether arrest the tempo of this cycle (in which every event is governed by the action of the preceding one) then it can no longer serve the legitimate

---

[8] Here the philosophical concept of 'culture' is not intended, but rather the forces of civilisation (always based on culture) as manifested in society and commerce. The closest approximation to the concept of culture applied here is when it is compared to Nature as an entity and a driving force. – V.S.

purpose for which, in common with all of Nature's creations, it is destined. The affected organism lags behind, thrust aside from the main evolutionary stream of Life. All those organisms whose life or death are dependent on it are also condemned to death, ultimately causing the demise of the foolish, interfering hand which is to blame for it all.

The causative force is our mind[9] and the soulless technology it has spawned, including our lawless and mind-destroying technological culture. These are jointly responsible for disrupting the circulation of the Earth's water and blood. Moreover, if everything this mechanistic civilisation has created should perish in step with such development, then the breakdown is in no way a passing crisis. It is the inevitable collapse of a dizzily-high, foundationless cultural edifice, whereby whatever is left of genuine culture will also be swept away.

Nature Protects Herself

Nature's most effective protection is the frailty of humanity, its work and its activities. The consequences of its activities must sooner or later bring about its own demise because the greater part of its present endeavours contravenes every principle of Nature. Hence, it is merely a question of the efficacy of humanity's activities and of the attainment of a particular level of culture which determines when the reaction sets in, when all that has been built up with so much care and sweat must once more collapse upon itself.

Once humanity has reached this point, Nature will rid herself effortlessly of her greatest enemy and with renewed energy will rebuild all that humanity has destroyed. If as a result, more and more people are to be found today who fear this fearful mayhem, they do it less out of a love of Nature than for their own self-preservation, which still remains a natural force in people. There are a few individuals possessed of great foresight, who are still in touch with Nature and are able to perceive the insanity of our work in its true light. Their ceaseless efforts are a solemn, though sadly unheeded admonition to their contemporaries. The latter are preoccupied with the exigencies of day-to-day existence and incapacitated by over-specialisation. They are no longer able to perceive the minute and fragile processes through which all Life in Nature is organically created and maintained, pulsebeat by pulsebeat. Unfortunately, the warnings to come to our senses are in the end but cries in the wilderness.

[9] Moving as it does in the realm of the phenomenal, the mind is unable to perceive the nature of the thing as such and therefore, as Kant says, remains limited in its categorising ability. – V.S.

The Laws of Nature
From *TAU*, No.153 (January 1937)

"Everything is governed by one law. A human being is a microcosmos, i.e. the laws prevailing in the cosmos also operate in the minutest space of the human being."
V.S. – (From *Implosion* Magazine, No.18, p.6)

Although governed by immutable laws, Nature is also subjected to eternal transformation. Within this constant metamorphosis lies hidden the profound secret of all evolution. Man alone clings to rigid and seemingly incontrovertible formulae and dogmas, and these are bringing about his undoing.

Our origin and our future are shrouded in an almost impenetrable secret that we are unable to unveil. That we cannot do so is because we do not believe in the eternal transmutation and reincarnation of all living things. We do not perceive that all development must cease if we fail to respect the inner stillness and seclusion of Mother Earth and continue to exploit the sunken residues of former life (coal, oil, *etc.*) for purposes other than those wise Nature ordained.

Nothing in Nature ever happens directly. One extreme always triggers off the other.

Nature's Rhythmical Processes
From *Implosion Magazine*, No.18[10]

Nature is not served by rigid laws, but by rhythmical, reciprocal processes. Nature uses none of the preconditions of the chemist or the physicist for the purposes of evolution. Nature excludes all fire for purposes of growth on principle; therefore all contemporary machines are unnatural and constructed according to false premises. Nature avails herself of the biodynamic form of motion, which provides *the biological prerequisite for the emergence of life*. Its purpose is to *ur*-procreate 'higher' conditions of matter out of the originally inferior raw materials. This gives the evolutionarily older, or the numerically greater rising generation, the possibility of a constant capacity to evolve. Without any growing and increasing reserves of energy there would be no evolution or development. This results first and foremost in the collapse of the so-called Law of the Conservation of Energy and consequently the Law of Gravity and all other dogma lose any rational or practical basis.

The solid, liquid, gaseous, etheric and energetic content of the organism – water, are not impurities as contemporary science would have us believe, but should be considered as aspiring (latent) energetic substances. With

[10] Extract from a V.S. letter to a certain Mr Kröger of Bochum in Germany, which was written in Vienna 30 Nov.1940. (*Implosion* Magazine, No.81, p.6) – Ed.

present methods of regulation these are expelled. They can only be retained in the water, if it moves sinuously along very particular systems of curves. The so-called carrying capacity and tractive force[11] of water are not derived from any mechanical thrust. They are the biological consequence of self-evolving life-forms, which are built up through a hitherto unknown form of tension or potential. As an energetic nucleus it is the product of the organic syntheses that occur after the atoms of the basic elements have been dissociated. The most interesting aspect in this regard is that the inner carrying and tractive forces in naturally flowing water can be separated, enabling their use as a natural, formative force or as a natural, motive force. The biological outcome of the exploitable inner force of this organism – water, on the one hand is almost unlimited *abundance of food*. and on the other, if the formative component of this natural force is switched off, absolute *freedom of motion* can be achieved. In short, the result is free and unlimited mechanical power.

Contemporary agricultural and economic practices are the cause of the decline in the quality of soil produce. From a more naturalesque[12] standpoint this concerns the suppression of all biological and indirect intermediate processes. In order to prevent this it is necessary to ensure that agricultural equipment for biological farming is suited to Nature's purposes. If we wish to continue to exist, we shall have to become accustomed to substantially different ways of working or adapt ourselves to natural forms of motion, whose patterns we can find in Nature. Then everything that we can see around us or are otherwise aware of will be provided in superfluity.

What I have stated above is absolutely no product of wild fantasy, but is a fact that can be proven in every case. No scientist can deny this, because the devices have already been built that produce these organic products of synthesis. With the aid of these 'higher' creative forces produced by biochemical motion, the fundamental Law of Mechanics (that resistance to motion increases with greater velocity) is rendered null and void. The resistance decreases with the naturalesque movement of the apparently lifeless contents of those organisms – air and water.

---

[11] 'Tractive force': This refers to the force described hydraulically as 'shear force' – the force that dredges and dislodges sediment. In German the term for shear force is 'Schubkraft', whereas Viktor Schauberger uses the word 'Schleppkraft'. The verb 'schleppen' means to drag, draw or pull. Viktor Schauberger's choice of 'Schleppkraft' here is quite specific, since in his view the movement of sediment is due to the sucking action of fast flowing, dense cold water downstream, rather than to the mechanical impact of the water coming from upstream. In view of this subtle change in emphasis, in lieu if the hydraulically correct term 'shear force', the term 'tractive force' will be used. This dynamic is similar to the effect of wind on roofs, where a roof is blown off not by force from the windward side, but rather by the sucking effect of vortices created on the leeward side. – Ed.

[12] 'Naturalesque': As adjective or adverb, in the Oxford English Dictionary this is defined as "Having the characteristics of Nature or natural objects." and "Imitation or adherence to nature." Its use here and elsewhere is to differentiate between processes and objects that occur naturally and similar objects and processes that are technically contrived so as to accord with Nature's own functions. – Ed.

Estranged from Nature, humanity does not understand water from which all life emerges. It believes that by abrogating water's rights and forcing it to flow according to its laws, it can build up the energies that evolve from the deceased remains of former life. These energies are necessary to allow the vastly increased, evolutionally older to come into existence so that after their own death they too can serve the following generations as a source of *'spiritual'* influx or *'in*-spire-*ation'*.

Nature Operates only Indirectly

"How else should it be done then?" is always the immediate question. The answer is simple: *Exactly in the opposite way that it is done today!* Very simple observations reveal that Nature's ways are always indirect. It is through our sheer intransigence that we always find it necessary to adopt the most direct approach. Therefore we should not complain if as a result we are constantly at odds with almighty Nature.

All we really need to do is adapt ourselves intelligently to Nature's marvellous order: To understand that it is indeed both foolish and futile to fight against her forces and to realise that if left to itself all would otherwise happen of its own accord. Moreover the recovery we so fervently desire would then come about quite automatically. Nature constantly indicates the right paths to take. Most certainly these new ways will lead in the opposite direction to the one we are wont to take. This is only to be expected, because it is the present direction that has led to our undoing. All those who seriously wish to travel this new road should take note of the following explanations.

Questions for Science

"My dear friends! We move everything back to front. What we are doing is incorrect and contrary to Nature. Nature moves in other ways. She primarily employs attracting or sucking energies, since these are indispensable to Nature for the growth and maintenance of life. Nature only uses pressural energies and explosive forces for destruction and reducing quality. The work of atomic physicists is also upside down. They would be more accurate if they started with simple nuclear fusion. They should set about the cold transformation of hydrogen into helium, as Nature has done over the millions of years of Creation. Today's technology has grasped the tiger by the tail, because it splits the heaviest atoms with the greatest development of heat and an enormous expenditure of energy."

V.S. – (*Implosion* Magazine, No.51, p.22).

Since the very beginning of time the Sun has stood above everything, staring down in icy silence at the frenzied activities of humanity, who regard it as a fiery orb. How could it be otherwise, such is their direct mental approach towards life? Yet the closer we approach this source of light and heat, the colder and darker its face will become. The nearer we are to it, the brighter the stars will be. As its light diminishes, heat, atmosphere, water and life will also disappear.

- What serves the Sun as a carrier of light and heat, if, in the view of our learned scientists, space is a vacuum?
- Why is the light and heat in the tropics more diffuse, and at the poles the Sun's light more intense and its radiant heat less?
- Why is water at the poles warmer at the bottom? Why is the sunlit surface so icily cold?
- Why doesn't the warmer, lighter bottom-water of the sea rise upwards?
- Why are water temperatures at the equator so warm? Why is it that it gets colder with increasing depth, and why does it get warmer again below the boundary layer of +4°C (+39.2°F), and why does Life begin there anew?
- Why do the magnetic lines of force run from south to north, and why does the Earth rotate from west to east?
- Why does a top stand upright when it is spun from the side?
- Why is the desert so dead despite all the heat?
- How is it that the warm Gulf Stream can push cold seawater aside and wend its way for thousands of kilometres over ocean mountains and valleys in a reversed temperature-gradient, without the assistance of a mechanical gradient?
- Why does groundwater in walls rise far above the surface of the ground?
- Why don't wooden posts rot under water, but always above it?
- Why do damp tiled roofs dry out from the eaves towards the ridge?
- Why can rising cold water pierce through the hardest rock?
- Why doesn't the Earth's warm air rise?
- Why is it so cold at the top of a mountain – nearer the Sun?
- Why is it warmer nearer the ceiling and colder at the floor in our houses when an artificial source of heat is used?
- Why do gases condense with a decrease in temperature, and why don't the fiery gases of the Sun, with supposed temperatures of over 6000°C (10,832°F), stream out into space?
- Why does marble expand with heat, and why doesn't it contract again with cold?
- Why do west→east flowing water-courses fertilise their banks?
- Why are the banks of east→west flowing rivers so barren?

- Why are the banks of south→north flowing watercourses fertile on one side only?[13]
- Why do rivers flowing into cold seas migrate laterally to the north?
- Why does the salt content of the seas vary?
- Why do herrings migrate northwards in winter?
- Why do deep-sea fish glow?
- Why do cold-blooded animals carry fever-inducing poison?
- Why does a cold fever occur in the tropics? Why does a warm fever arise from a chill? What is fever anyway?
- What is temperature? What is heat? What is cold?
- What is energy?
- Why does the heart beat in our breast? Who gives this muscle its impulse to move? Where is the motor for this pump? Why does blood circulate in our blood vessels? Why do we breathe day and night, when asleep and even when totally unconscious?
- Why do the fluids in a chicken's egg circulate without a heart? Why does a stone suffocate when we cut off its air supply?
- Why do light-demanding timbers have a thick bark, and shade-demanders only a thin one?
- Why does a trout stand still in a raging torrent, as if by magic?
- What is it that keeps the Earth floating in space?
- Does the heart beat because we breathe, or do we breathe because the heart beats? Where is the heart of a plant?
- Why does water pulsate and breathe? Why does groundwater manage to remain on the sides of mountains and why, growing colder and heavier, does it rise upwards? Why does it frequently spring from high peaks?
- Why do deltas and estuaries develop?
- What is evaporation? What is vaporisation?
- What is dissolution, what is combination, what is absorption, and on what effects are these processes founded?
- Why is our body-temperature sub-normal when climbing a mountain and above normal as we descend?

The Error of Civilisation

Is there really such an enormous difference between the breaching of a riverbank and the bursting of our blood vessels? Is it really necessary that the last human being must rot away alive before we all become consciously aware of the errors of our ways? Why can we not admit to ourselves that it

---

[13] North->south and south->north effects are reversed in the southern hemisphere. – Ed.

is our senseless activity that is killing us? Do we actually have the right to stuff such worthless knowledge into our children, when science has already led us to the very brink of disaster? Where does our knowledge begin and where does it end?

Does anyone still dare to speak of science and culture in the same breath? Are our children actually wrong if they refuse to be instructed by their parents and teachers, and choose to go their own way? Is one seriously to believe that hunger can be appeased by political phrases and bayonets? Are there really still people who believe that improvements can be achieved through coercion, when work undertaken of their own free will has already brought such unspeakable misfortune?

If this be true, then let Nature quietly continue to so prevail, for she will then do great and noble work. Nature is simpler in her effects and more complex in her functions than our rational minds can conceive.

The Road To Free Energy

More energy is encapsulated in every drop of good springwater than an average-sized power station is presently able to produce. These energies can be generated effortlessly and almost free of cost if we follow the path which Nature constantly shows us and abandon the blind alleys of conventional technology.

Happiness and health are available to us just as near cost-free as unlimited energy, if we but once realise that in water dwell Will and its resistance, Life. We struggle so hard for these today, because in all our endeavours we constantly rob the bearer of all Life (water) of its noblest possession, its soul. The Will of Nature serves all things and expresses itself in growth by way of atomic dissociation and transformation. It is only through our obsession with atom-destroying work and our selfish over-exploitation of her resources that we encounter Nature's resistance.

The only possible outcome of the purely categorising *compart*-mentality, thrust upon us at school, is the loss of our creativity. People are losing their individuality, their ability to see things as they really are, and thus their connection with Nature. We are fast approaching a state of equilibrium *impossible* in Nature. This equilibrium must force us into total economic collapse, for no stable system of equilibrium exists. The principles upon which our actions are founded are therefore invalid because they operate within parameters that do not exist.

Our work is the embodiment of our will. The spiritual manifestation of this work is its effect. When such work is properly done it brings happiness, and when carried out incorrectly it assuredly brings misery. Humanity!

Your will is paramount! You can command Nature if you but *obey* her. Do not complain if you must become her slave!

Concerning Micro-Organisms

Let us look at some situations in which bacteria are produced. House floors were traditionally constructed of softwood such as pine or fir. These were frequently washed and lasted for decades, even when the gravel underneath was permanently wet. As new styles of interior decoration became fashionable, people wanted hardwood parquet flooring, which was laid directly over the softwood sub-floor. When these parquet floors were washed, micro-organisms sometimes developed and multiplied so that the superimposed flooring disintegrated in the space of a few years. In such cases our experts maintain that the timber employed was already infected. The true facts are substantially different. The structure of fine hardwood is of a higher quality than the more coarsely-grained softwood. Fine-grained wood contains qualitatively higher-grade proteins which metabolise only very slowly with a normal supply of oxygen.

If there is sufficient space between the old lower floor and the parquet floor, so that no air-tight intermediate layer can develop between the materially different types of wooden flooring, then these floors can last for decades, provided the wood is of suitable quality. If however the upper floor is washed and the intervening space is sealed off as the wood swells, then between both floors a warm, humid layer is created. Owing to defective water-proofing this now obtains its air and oxygen supply from the rising groundwater in the walls. This is water that has not been exposed to the Sun.

The concentrated oxygen rising with the uninsolated groundwater will expand in this moist, warm, intermediate zone and become aggressive. In this condition, this highly-organised oxygen first of all combines with the less-complex proteins of the sub-floor. The energies resulting from these metabolic processes provide the impulse for the development of certain micro-organisms, which begin their vital activity at appropriate ambient temperatures and eat away the parquet floor from the bottom upwards. Different types of food and micro-climate propagate different strains of these micro-organisms. They eventually infest the wider environment once their original breeding ground has been destroyed. It is therefore obvious that sickening trees in the forest will also be invaded by parasites. This especially affects shade-demanding species planted by modern foresters in the open; their sap becomes highly oxygenated and exhibits a much coarser structure. These phenomena are only the secondary and subsidiary

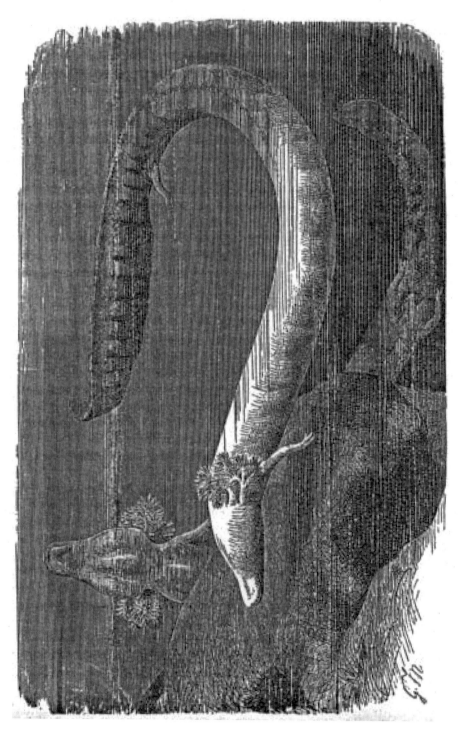

Fig. 1: A Grotto Olm.

after-effects of clear-felling operations practised over the last hundred years or so. The primary cause of the serious damage ensuing from clear-felling will be addressed later.[14]

The question of how the microbe world comes into being and all its various preconditions cannot be usefully addressed as long as water continues to be viewed as a lifeless substance and its inner metabolic processes are not taken into account. It is always the incessant metabolic activity of and within water that generates a particular life-form. Regardless of whether it is beneficial or harmful to humanity this life-form will ultimately serve the build-up of the whole.

Another instructive example concerns the living conditions of the grotto-olm, a blind, cave-dwelling salamander of the genus *proteus* (fig. 1). If we study the water in subterranean lakes, where the influence of light is totally excluded, then in such water we find an extremely peculiar atmosphere and no microbial activity. Apart from olms, present in these lakes in great number, there are no other living things. On what, then, do the olms live?

The heavy concentration of oxygen in such water requires only a slight warming and a consequent increase in aggressiveness in order to transform highly complex carbones[15] into an even higher quality, which the olm then ingests with the atmosphere contained in the water. The olm's respiratory processes and bodily heat trigger strong oxidising phenomena, leading to the development of increased heat. Together these are sufficient to transform highly complex carbones in the olm's body into the kind of food it requires to sustain life.

On the other hand if the olm is removed from the cave and exposed to increased oxygen, the surface of its body begins to discolour and the olm dies. However if the olm is immediately placed in a container at the place where it was caught and if it is not exposed to the light of day, and if warm rainwater is poured into the container, then the identical phenomena occur. Again we encounter the same principle which for example also explains the mountain trout's peaceful stance amidst rushing water.

The above examples, however, are insufficient to clarify the true nature of autogenesis (spontaneous generation), which was acknowledged in the Middle Ages, but is rejected in modern times. Another simple example

---

[14] These phenomena, damage to timber and underfloor decay, are discussed in greater detail in The Fertile Earth – Vol. 3 of the Ecotechnology series. – Ed.

[15] 'Carbones': In contrast to the normal use and definition of 'carbon', Viktor Schauberger grouped all the known elements and their compounds, with the exception of oxygen and hydrogen, under the general classification of 'Mother Substances', which he described with the word 'Kohle-stoffe', normally spelt 'Kohlenstoffe' and meaning carbon. Apart from the above definition the hyphen also signifies a higher aspect of carbon, both physically and energetically or immaterially. The additional 'e' in the English word is therefore intended to redefine and enlarge the scope of the usual term 'carbon' in accordance with Viktor's concepts. – Ed.

brings us even closer to the facts of the matter. Those places where very dark and glistening water streams out of the Earth's surface are the spawning-grounds of fish for good reason. If we examine such water at the very limit of light penetration (at the place where it first encounters incident light) a noticeable change can be detected in the matter found in such water and, the first beginnings of bacterial life. The closer we approach the zone shielded from light, the more highly evolved the bacterial life in the water becomes. Conversely bacteria are increasingly less complex the longer water flows in the light.

If we observe the fish living there the same picture emerges. The closer the fish to the spring, the tastier they are. Every fisherman knows that the powerful, stationary trout which live close to the spring, spurn every type of lure. Another remarkable fact is that these fish can live for months in caverns, to which they migrate when the water subsides during the hot summer months. The feeding habits of these creatures, which spend half their lives in daylight and half underground, are substantially different to those of fish living in the lower reaches of rivers and are similar to the lifestyle of the olm. A fact well known to alpine hunters is that the consumption of these almost-blind fish leads to higher sexual potency.

Another very interesting phenomenon is demonstrated by the emergence of mealworms. If a vessel containing meal is placed in a dry, warm spot, then only a few worms come alive, or none at all. In order to obtain a greater quantity and better quality of worms, an old woollen cloth or a bone is placed in the meal and the lid closed. The increase in the worm population is caused by the introduced third category of carbone whose origin stems from a group of vegetable matter more complex than the meal.[16]

While on the subject, a few interesting experiments can be described. If we pour a dilute solution of potassium chromate, or iron or copper sulphate, onto a moist gelatinous film, beautiful patterns of deliquescence appear which under the magnifying glass exhibit a delicate, strongly branched structure. If river water is used to make the gelatine and the whole experimental arrangement is placed at the interface between a positive and a negative temperature gradient, after a certain time various fungi, algae and mosses can be detected under the microscope. On the other hand if fresh seawater is used instead of fresh water, then different flora and fauna of this microbe kingdom will appear which are characterised by more worm-like, wriggling organisms. Under the right conditions this microbial world behaves in the same way as its brothers and sisters in the

---

[16] Though not otherwise clarified, this would seem to infer a higher form of carbone evolving through the consumption of vegetable matter. – Ed.

macro-world. It devours everything around it while engaging in the mutual struggle for existence, excreting all unusable matter and reproducing itself with incredible speed. The results of this experiment are particularly clear if the procedure is carried out in a well sealed glass vessel, insulated externally in order to maintain the correct temperature gradient (in this instance created artificially) and to prevent any transfer of energy to the outside. Apart from an atmosphere conducive to the formation of the desired micro-organisms or worms, the presence of a *third, more highly-organised substance* is necessary in order to activate the energies and to create the conditions: for example, a drop of oil in water of the appropriate composition.

Whether the propagation of these micro-organisms is caused by their own physical energy or through the effect of an artificially-created temperature gradient, is quite immaterial. The most important factor in both cases is the necessary alternation of the climate over short periods of time. This frees the life-generating energy at the point of intersection between the individual climatic zones (at the interface between two complementary temperature-gradients). Once again the prerequisite for the success of this experiment is the correct proportion of basic elements, the oxygen and carbone groups contained in the water. An appropriately-shaped vessel with a suitable air-tight seal is also required, in which the inner climate can be produced and maintained which is suited to the respective creature and therefore necessary for its life activity.

Another example can be mentioned to clarify a natural phenomenon which science has so far been unable to explain, but which can be clarified easily if we observe the circumstances that give rise to these remarkable events. These are the so-called 'worm rains' in Lapland which now and then happen in spring, during which it rains living white worms about 3cm (1") long. The usual, but incorrect assumption is that these worms, which fall from the heavens under the *blood-red light* of the midnight Sun, are somehow, somewhere caught up by wind, gathered together into a worm-cloud and at a particular location fall back to Earth in their thousands.

A similar curious phenomenon is the so-called 'rotting season' that starts in Lapland towards the end of July and lasts about four weeks. Trees should not be felled during this period because after a few days fungus appears in such profusion that all the work is in vain. Even heavily salted bacon is tinged with all manner of colours. The smallest wound festers and can only heal after the rotting season has passed. The same applies to animals, since any wounds they suffer during this period are also incurable. The young born in this period of purulence are often deformed. Mosquitoes and other pests die off *en masse* after the rotting season is over.

Further proof that a certain seasonally dependent climate or *particular light influences* enhance the propagation of a superabundance of micro-organisms is provided by epidemics that regularly occur under certain pre-conditions. These are caused by bacteria alone. They represent Nature's most effective self-defence when the human organism foolishly interferes with the driving forces of Life and Nature.

It is well-known that varying intensities and qualities of sunlight peculiar to each season play a major role in growth. For example, if light is directed into a room through window-panes of a particular colour then the flies begin to die off. However, if the colour of the glass is changed they can revive. The decrease in tuberculosis since radio waves first vibrated through the ether is also no accident. These emissions cause an unbalanced and excessive concentration of oxygen both in water and the atmosphere which goes a long way towards explaining why human beings have become faster-living, more hot-tempered, but regrettably less intelligent in the process.

Studies of earth rays and the appearance of symptoms of cancerous decay frequently associated with them, reveal that these can also be traced back to interactive processes in the interior of the Earth which have been impaired locally. These have been unfavourably affected by shifts in the distribution of groups of basic elements, for which groundwater acts as a conveyor, distributing them to all life via the soil's capillaries. All these phenomena, so mystifying to science, can be duplicated or prevented once the nature of the primal substance of all life, *the nature of water*, is understood.

Water in Ritual, in Life and in Medicine

The following section is about the deeper implications of water in the life and ritual of our ancestors, and takes the form of an investigation of historical symbolism, borrowed in part from the works of Martin Ninck, Norden, Weinhold and others.

Having little or no time for inner composure, or the contemplation of human developmental history, modern people naturally see water as a purely chemical substance, adequate for our physical needs as bathing water and possessing a purely practical value as the driving force for our power stations. Our ancestors viewed water from a completely different standpoint, seeing it as the source of all life. Many legends and transmissions from the mythology of various earlier peoples conceal a much more profound meaning than is usually ascribed to them by their more rational but less deep-thinking descendants.

The point of view expressed in my explanations, that water is to be considered the blood of the Earth, finds its parallel at many points in our ancestor-worship[17]. Various sayings and depictions make reference to mother's blood, mother's milk and the maternal tears of our ancient Mother Earth. Even modern linguistics owe much to the symbolism of earlier epochs. It is therefore no accident that the word 'spring' has a feminine connotation.[18] The figures of the water-goddesses, the nymphs, are always coupled with stories of love. Nymphs are ready at all times to give birth, as Goethe also said of the spring in Faust:

"For a spring abrim with songs of love is constantly reborn."

Wuttke-Meyer also cites the following old German custom in connection with the fertility of springs. When going to a spring for the first time, every pregnant woman had to 'silver' it by throwing in a coin, otherwise it would dry up. Apart from the springs, rivers and lakes were also highly venerated in the rituals of the ancients. Even today we find that the distinguishing characteristics of the principal rivers are allegorically portrayed by their tutelary deities. According to whether the water is in motion or at rest, it will be ascribed either a male or female fertility potency. In *The History of Religion* by Chant de la Sauss, we discover that the ancient Egyptians believed the ur-water Nun to be possessed of a dual potency.

In his lyrical ode, 'God, Nature and Cosmos', Goethe writes:

There, where water splits in twain,
Life is e'er set free, unfolding its domain,
And in emerging from its source,
Water's blessed with vital, living force.
There flock beasts, a-thirst for flowers,
'Midst thrusting boughs and leafy bowers.

And in Faust, the same German prince of poetry declared:

You sources of all life,
upon whom hang Heaven and Earth,
you spring forth, you overflow!

---

[17] Schauberger wrote elsewhere (*TAU*, 137, p.16.): *"Similar things are reported by the priests of antiquity. In their rivers, such as the Nile, they constructed sluices in the form of temples. The religious populace believed it was the Gods who brought fertility to their fields in such a manner. In reality it was the knowledgeable and shrewd high-priests of the waters, who in this simple fashion maintained and looked after the primordial energy, the ur-life of the water, from which the legendary fertility and proverbial prosperity developed."* – Ed.

[18] In German, the word 'spring' *Quelle*, is preceded by the feminine article (*die*), and is therefore con-strued as being of feminine nature. The English language unfortunately makes no such differentia-tions; there are neither feminine nor masculine nouns. While in no way denying the necessity for redressing the long-standing imbalance between the sexes, this lack of differention in English may in some way have contributed to the apparently greater force of the feminist movement in the English-speaking world than in countries, where the natural male-female balance is to a large extent provided for and supported by the structure of the language itself. – Ed.

In order to explain the importance of water in medicine, it would be best to permit a doctor still connected with Nature to speak for himself. Dr Schew writes:

"In the nature of things, water is the great bestower of energy. It is the most invigorating, and at the same time, the most powerful of all tonics. In this regard, there is nothing else like it in the whole world."

In his book on natural methods of healing, F. E. Bilz lets the great poet himself express this point:

"This vast expanse of water – the ocean – is the condensed breath of God, without which all would be but a cold and barren mass of rock. It is a breath that has endowed the Earth with fertility, beauty and life."

The role played by water in the constitution of the human body resides in the fact that the body consists of up to 90% water. It is much easier for human beings and animals to go without food for a long period than to go without water. The average person can survive for about three weeks without food and water. However, if water containing a certain quantity of nutrients in material and energetic form is drunk, then such a person can last for considerably longer. A Dr McNaughton tells of a madman who was able to survive for 53 days on water alone.

Today, modern civilised people drink predominantly bad water. As a result they have to a large extent given up drinking any water at all, thereby inflicting serious damage on the body. Dr Munde writes:

"Recent investigations by Genth, Bequerel and others reveal that an increased discharge of moults follows from an increased intake of water in the body, whereas a reduced intake of water results in a greater condensation of the same, and a greater quantity of uric acid in the urine – a fact of which those stricken with gout should take heed. As can be determined by comparing various medical experiments, there is a certain optimum quantity of water for every individual, which very significantly raises the quota of solid matter in the urine."

Finally, attention should be drawn to the fact that people who consistently drink good, healthy water also have a good appetite and consequently probably stay healthier.

Our Senseless Toil – Conclusion

While the previous explanations may have been couched in rather harsh terms,[19] this is done in the public interest. The danger will increase daily should present methods continue to be applied. If determined action is not

---

[19] Due to the many unhappy experiences he suffered at the hands of Academia, Viktor Schauberger tended to be extremely trenchant in his critique of their behaviour. While his derogatory remarks may appear repetitive, though from his standpoint quite justified, in order to maintain the integrity of the following articles as he wrote them, these comments will be included. – Ed.

taken quickly then chaotic conditions will inevitably arise within a very short space of time. There is absolutely no time to be lost.

It is understandable that our hydraulic experts take no delight in the ruthless revelation of this deplorable state of affairs. However, this does not alter the facts of the matter. Numerous attempts in recent times to withhold my many informative dissertations from the public by removing relevant articles from books, newspapers and periodicals are childish. By so doing, they will ensure that I make even greater efforts to place these expositions before the public. Such attempts only demonstrate a certain weakness, and in any case are merely proof of the irrefutability of what I have stated. Frequent protestations that 'contemporary methods are being practised world-wide and hence cannot be wrong' mean absolutely nothing. At best, they serve only to explain why the whole world is in such an appalling state. In this situation the best strategy can only be a ruthless attack, in which all parties are at all times entitled to the right of defence and rebuttal. All those who wish to perpetuate the status quo for fear of losing their jobs should bear in mind that an existence built on false foundations cannot in the long run be maintained. That is, even when it appears guaranteed by security of tenure, because an impoverished people pays no taxes and therefore cannot afford an expensive bureaucracy.

When they become ill, let those who continue to believe that water is a lifeless substance to be controlled by mathematical formulae alone, then summon a mathematical genius to their sickbed instead of a doctor. Their fellow men will then be rid of such narrow-minded thinkers as fast as possible.

Unfortunately the catchword 'systematisation', which all too frequently governs our methods of working nowadays, has found greater favour than is good. Today, the expression 'logical thinking' or 'mathematically-trained thought' conceals sheer intellectual incapacity or mental inertia. By far the largest number of discoveries and inventions have not been made on the paths trodden by scientists, often to their astonishment, if not to their great dismay. The overall progress of the world is caused by a certain measure of discontent, the characteristic phases of which are revolutions or wars. Similarly great advances in the realm of the intellect are brought about by revolutionary thinkers.

A certain myopia has gained ground even in the empirical methods practised in contemporary hydraulic research institutes. People still cling desperately to the external appearance of a given phenomenon, thereby failing to study the far more important nature of its inner processes. Indeed, amongst those in this field who are responsible there are some who have already recognised the limited value of such purely superficial observations. Yet, for reasons of job security, they regrettably continue to abide by tradition.

There are many greedy individuals who believe that water, oil, coal and other precious substances can be torn from the Earth with impunity. Concerned only for their own well-being, they are quite prepared to create a black market not only in foods, but even in the water destined for rich and poor alike. Such persons should be warned that the despair of the great mass of the people will bring about a much earlier end to their selfish endeavours than they could ever foresee. Everyone else, especially our young people, should cooperate as a first priority in the rebuilding of our former indigenous forest, and with this the restoration of healthy water to the Earth. Then we shall all be able to survive and humanity will continue to exist.

This year (1933) is to be remembered as the hundredth anniversary of Alfred Nobel (1833-1896), the Swedish chemist and engineer, who became a millionaire through the invention of dynamite in 1866. Becoming aware of the tragic consequences of his invention a few years before his death, he established the Nobel Foundation, no doubt in the desire to make amends for the frightful harm he had caused. Millions of human lives have since fallen victim to this fearful device of destruction and war. Further millions, perhaps even whole races, will yet be robbed of their lives, of their very existence, if humanity continues to avail itself of such inventions. It is not unique for humanity to have to traverse tortuous and often dangerous routes before arriving at a better understanding of a matter.

If we study Gilbhart's article, which appeared in the business section of the *Deutsche Zeitung* (no.242, 1933), we see that the German government is now in the process of rectifying a serious mistake committed about a hundred years ago. It caused perhaps even greater harm than the use of explosives in modern warfare. This article was entitled 'New Forestry'. Scientists have long been aware that many practices in modern forestry are unsound and that our forests have declined in quality since the introduction of scientific forestry methods in about the middle of the last century. Yet they have so far failed to show the necessary courage to own up manfully to their mistakes. With the prohibition against clear-felling much has been achieved, but not all that is necessary. This was done merely to gain time in order to safeguard their employment, or at least to make provision for their old age at the expense of the general public. It will not succeed however, because a terrible evil has been foisted on the entire human race by forestry's supposed science, will be studied in such detail that every schoolchild will understand the result of its ignorance of the true facts. The same applies equally to modern agriculture and other contemporary achievements.

Considering armaments and preparations for war, it is possible to render harmless all bombers, dirigible gas balloons and even gas and explosive filled grenades by the simplest of means. All other weapons of war will become like children's toys once humanity fully understands the energies that slumber in water. For this reason, equal care will also have to be taken to ensure that everyone learns how to use these forces. For if humanity is truly intent on self-destruction through the use of force, then it should have the requisite means placed at its disposal, so that it can fulfil this desire as quickly as possible.

This should leave the reader with mixed feelings. It should be quite apparent that, in view of many undoubtedly correct observations and well-founded clues, it should not be necessary to attack science and technology as fiercely as they have been attacked here. However, because it would be utterly futile to seek any other worthwhile contact in this direction, unfortunately, attack is absolutely necessary.

The following explanations, which will be elaborated further in later publications, should furnish sufficient proof of the impossibility of a compromise or of the incorporation into the existing scientific edifice of discoveries only touched upon here. The errors of contemporary science and the damage wrought by today's technology are too great. With the present (1933) consumption of coal and oil at a level of two thousand million tonnes per annum, the moment when this important source of energy will be exhausted is coming disturbingly closer. Within a few centuries, as science has calculated, the last reserves of oil will have been extracted from the Earth. If we continue to busy ourselves in the same old way, we will have to search for other sources of energy since the loss of these energies would signal the end of our civilisation.

Science earnestly endeavours to discover new forms of energy and also believes these can be obtained through research into cosmic energies. However this pursuit demonstrates not only an almost boundless bias in thinking, but also furnishes irrefutable proof of the untenability of scientific endeavours and objectives. To put it mildly, they can only be described as Utopian. A science possessed of such goals cannot possibly be taken seriously, and cannot claim the right to a leading role in the fate of humanity.

The direction indicated earlier is far from being the only correct one. Nevertheless it is closer to the truth and is thus able to throw a revealing light on these misconceptions, since the ascertainment of the purest truth is an unattainable goal for humanity and will probably ever remain so. The very idea of using alternative (cosmic) energies once all the reserves of coal, oil and timber have been exhausted is so absurd that this alone condemns the whole of science.

Temperatures prevailing in the interior of the Earth are the product of interactions that take place between carbones contained in the Earth and oxygen entrained by infiltrating water. Were the last reserves of highly-organised carbones eventually to be totally stripped from the Earth, these interactive processes could no longer take place and the Earth would cool off. Since it is practically impossible to remove all carbones from the Earth however, these cooling phenomena can only occur to an extent commensurate with the severity of disturbance to these inner interactions. These disturbances are caused by the removal of carbones, or by ventilating the Earth. Conforming to natural law, the effects of today's technological and industrial intrusions into the Earth must therefore lead to the following results:

If various external influences such as bore-holes, deep wells, shafts and open-cut mines, excessive extraction of coal, metals and minerals, all act to inhibit these interactions, they will provoke a cooling of the Earth's crust. The atmosphere will also cool off as a further consequence. These causes, which ensure an excessive accumulation of oxygen in the atmosphere, must also result in its concentration due to the influence of cold. In the course of time air strata normally subjected to low atmospheric pressure will become heavier in the absence of upward-streaming groups of carbones. These strata will sink downwards, over-saturating both the vaporous and the fluid hydrosphere with oxygen. If, having now become over-saturated with oxygen and heavy, this water succeeds in infiltrating into deeper strata of the geosphere, into the carbonesphere for example, then under the prevailing high temperatures the accompanying oxygen will trigger off lively oxidative events.

The cumulative effect of these will result initially in localised explosions, eruptions or earthquakes, and subsequently in ruptures of the Earth's surface. This will provoke the sudden release and elevation of gaseous carbone groups. These relatively elementary substances will first interact with atmospheric oxygen only at great altitudes, and in various hot zones will trigger off a regional redistribution, causing sudden cold spells and the movement of stronger or weaker air currents.

In equatorial zones the ascent of these carbones will be enhanced by the stronger reflection of heat. Under certain circumstances the reciprocal effects thus provoked can become so large and so aggressive that the zone of interaction in the lower levels of the atmosphere extends downwards in the form of funnel-shaped clouds. This leads to the general formation of tornadoes and the violent cyclonic storms which have long been known in equatorial regions. Through these powerful interactions, water-vapour will be forced into localised concentrations, resulting in the formation of heavy thunderstorms and the occurrence of cloudbursts. Apart from large quantities of carbones, substantial amounts of water-vapour are also expelled into the atmosphere during strong eruptions, offering increased resistance to the Sun's radiant energy, and hence to an increase in heat (through the heat-absorbing function of atmospheric water-vapour).

The sequel to this phenomenon is a short-lived, luxuriant profusion of vegetation (a phoney agricultural success) in which consumption of gaseous carbones (carbon dioxide) is intensified. These substances, however, can no longer be produced from the Earth's interior in the necessary and regular proportions. This leads to a qualitative decline in various forms of vegetation, and to a decomposition of the dynagens[20] reflected back by the Sun, namely to a systematic cooling-off, and therefore to the inauguration of a *new ice age*.[21] These developments will soon be brought about by the devastating activities of those involved in forestry, agriculture, water and energy-resources management because, as a result of their predominantly one-sided way of looking at things, the regularity of the water cycle will be inhibited, and with it the energy cycle and the upward flow of carbones. As already mentioned, it is inevitable that humanity's present absurd practices will bring about a drop in the quality of dynagens reflected back by the Sun. Ultimately, by arresting the oxidative processes in the atmosphere, the generation of heat will also be reduced.

It can thus reasonably be asserted that the next ice age will be virtually dragged in by contemporary science and technology. For this reason the manifestations of economic decline, familiar the whole world over, must logically keep pace with the advance of technology. This state of affairs will worsen at the same tempo as the sources of energy required to maintain technological progress are removed from the Earth. The greater the progress we achieve in technology, the deeper we will and must sink culturally and economically. However, this is not the end of the matter!! With the curtailment of the absolutely essential oxidative processes in the Earth, enormous amounts of water must make their appearance, initially in the atmosphere and subsequently in the Earth, because in neither case can it be assimilated or reconstituted. Now heavily over-saturated with oxygen and poor in carbones, this water, which either infiltrates into the Earth or quickly re-evaporates and streams upwards into the atmosphere, will reach its freezing point due to the absence of its partner, the carbones. This results in an unavoidable and fundamental change in general climatic conditions.

Furthermore, when the potential of the groundwater, now possessing a uni-polar charge, has been reduced though lack of carbones, it is forced to

[20] 'Dynagen': Belongs to the higher formations of energy described elsewhere as *Ethericities*: The term 'ethericity' refers to a supra-normal, near non-dimensional, energetic, bio-electric, bio-magnetic, catalytic, high-frequency, vibratory, super-potent entity of quasi-material, quasi-etheric nature belonging to the 4th and 5th dimensions of being. As such it can be further categorised as a fructigen, qualigen and dynagen, which respectively represent those subtle energies, whose function is the enhancement of fecundity or fructification (fructigen), the generation of quality (qualigen) and the amplification of immaterial energy (dynagen). According to their function or location these may be male or female in nature. – Ed.

[21] Readers are reminded that all this was written by Schauberger long before science had the vaguest inkling of such phenomena as 'global warming' and the 'greenhouse effect'. – Ed

sink down to depths where carbones may eventually still be present. There it shifts its boiling point, oxidises prematurely and gives rise to violent eruptions. With the final subsidence of the water, all vegetation will gradually disappear in the same way that it once appeared. After the occurrence of immense catastrophes, which will manifest themselves in the form of earthquakes, cloudbursts, whirlwinds and so on, the vegetation zone, in conformity with natural law, will slowly but surely *be covered with ice*. Deluges and catastrophic inundations are already on the increase everywhere today, to which approximately 20 million human lives will fall victim. Presently these are only harmless events compared to the disasters which can be expected in the future. These must inevitably occur if humanity continues to allow itself to be guided and controlled by contemporary science.

As a case in point, the present explanation for the formation of rain is so incomplete that it is hard to believe how such a hypothesis could have been upheld for so many centuries. Science explains the origin of rain as the condensation of atmospheric water-vapour due to the presence of cold. This explanation is approximately true in its widest sense, but in the final analysis it actually describes only a very unimportant secondary effect. Even the formation of rain, as elementary experiments prove, must be attributed primarily to the above interactions. These can only arise when ascending carbones intersect with descending oxygen groups attached to microscopic particles of ice.[22]

Were the scientific view correct, then in the higher strata of the atmosphere it ought to rain in winter and snow in summer, since it is well known that air and ground temperatures swap places with each other with the alternation of the seasons.

In consideration of the events described only in broad outline here, there is only one practical option left. That is to make humanity either suspicious or rebellious. Only thus, at the eleventh hour, will it still be possible, perhaps, to trigger off the necessary impulse towards recognition of the unimaginable danger that exists. This is a danger which today threatens a hoodwinked humanity regardless of race or nation. There are people, endowed with an almost God-given cluelessness, who have brought us to this terrible impasse and would lead us on into chaos also. Those apart, there ought to be people with enough humanitarian sensitivity to prevent our children from being led unawares into such frightful cataclysms. The possibility still remains to rip the blindfold from the eyes of sensible individuals and with their help, to undertake a rescue attempt. Our young people would doubtless energetically support this because it is their future that is at stake. There is no herbal remedy for stupidity, and therefore the unconsciously insane can hardly be

---

[22] See "The Learned Scientist and the Star in the Hailstone" in *The Water Wizard*, vol. 1 of the Ecotechnology series.- Ed.

called to account. However, if the causes of decline appearing everywhere are recognised for what they really are, then any continuation of these perverse practices, which are consciously leading all humanity towards disaster, would unquestionably be deemed a criminal act.

Conforming to natural law, the systematic disturbance of the water balance will reliably happen as a consequence of current industrial, technological and economic practices, leading to the increasingly extensive suppression of interactions that condition all life in Nature. The logical outcome of the cessation of oxidative processes occurring between basic formative substances is an increasingly widespread cooling and desolation of the all-nourishing vegetation zone. If the present *modi operandi* continue to be applied, then apart from the emergence of disease and degeneration, worldwide famine must inevitably follow.

The insights gained from the preceding explanations must force us to a decision. We renounce the grievously damaging achievements of contemporary science and technology and we strip our reigning intellectual masters of their power. Alternatively we allow them, little by little, to strip us of ours, and by every trick in the book to put us on ice (in the true meaning of the term). In this scientifically contrived, conserved form, we will be preserved for as long as conceivably possible. At the very least we will eventually provide a future humanity with the cautionary end-product of a bygone 'culture'.

Vienna, November 1933.

Let The Upheaval Begin!
From *Implosion* Magazine, No.67.

Even in earliest youth my fondest desire was to understand Nature and through such understanding to come closer to truth; a truth I was unable to discover either at school or in church. In this quest I was thus drawn time and time again into the forest. I could sit for hours on end and watch the water flowing by without ever becoming tired or bored. At the time I was still unaware that in water lay hidden the greatest secret. Nor did I know that water was the carrier of life or the *ur*-source of what we call consciousness. Without any preconceptions, I simply let my gaze fall on the water as it flowed past. Only years later did I come to realise that running water attracts our consciousness like a magnet and draws a small part of it along in its wake. It is a force that can act so powerfully that one temporarily loses consciousness and involuntarily falls asleep.

Gradually I began to play a game with water's secret powers; surrendering my free consciousness and allowing the water to take possession of it for a while. Little by little this game turned into a profoundly earnest venture, because I realised that one could detach one's own consciousness from the body and attach it to that of the water. When my own consciousness eventually returned, the water's most deeply concealed psyche often revealed the most extraordinary things to me. As a result of this investigation a researcher was born who could dispatch his consciousness on a voyage of discovery. I was thus able to experience things that had escaped other people's notice, because they were unaware that a human being is able to send forth his free consciousness into those places the eyes cannot see. By practising this blindfold vision, I eventually developed a bond with mysterious Nature, whose essential being I slowly learnt to perceive and understand.

In the process it gradually became more and more apparent that we human beings are accustomed to see everything upside down and back-to-front. The greatest surprise was the way we allow our most valuable asset, this great spiritual gift, to flow away unused and retain only the rubbish. This wasteful tendency we owe principally to books that were originally the product of an intuition, but which subsequently became merely the repositories of residual understanding. Books seldom had any connection with their original spirituality. As a result I became an enemy of the so-called erudite, the scientists and scholars, who set down their intellectual dross in black and white. The unsuspecting masses had to accept this garbage as gospel truth and became perverted by it.

As a huntsman I gradually developed a great respect for the instinct of supposedly inferior creatures. In their close connection with Nature, they manifest the unadulterated thoughts that Nature has impressed into their animal brains. It became clear to me that consciousness, which has been given to humanity as the final link in Nature's evolving chain, represents a great danger to a capriciously-minded humanity and its habitat. Free consciousness is least free as long as anyone attuned to Nature fails to understand how to make intelligent use of this divine gift.

Everything in Nature is built up sequentially, step by step. Therefore as a rule the intellect is something of an outlaw. It runs the gauntlet of a thousand dangers and is subjected to perils more hazardous than unsuspecting humans can imagine. It is therefore inevitable that so much deceit abounds in our societies and that humanity is sinking deeper and deeper into a mire of depravity. This is due to the fact that humanity followed the false dictates propounded by our great philosophers, who then speculatively translated their pseudo-knowledge into a logic generally referred to as mathematics. Kant made the fallacious assertion that any knowledge to which mathematics cannot be applied cannot be described as science. Instead of becoming a creator on Earth, humanity became a vulgar speculator, more and more estranged from truth and reality. Consequently, we lost all connection with Nature and became outlaws in the truest sense of the word.

Humanity has become accustomed to relate everything to itself (anthropocentrism). In the process we have failed to see that real truth is a slippery thing upon which the perpetually reformulating mind passes judgement almost imperceptibly. In the main all that is then left behind is whatever was drilled into our brain with much trouble and effort, and to which we cling. To give rein to free thought, to allow our minds to flow freely and unimpeded, is too fraught with complications. For this reason the activity arising from these notions inevitably becomes a traffic in excreta that stinks to high heaven, because its foundations were already decayed and rotten from the very beginning. It is no wonder, therefore, that everywhere everything is going wrong. Truth only resides in all-knowing Nature.

In the course of all this, the mass of humanity unwittingly became the advocate of a pseudo-knowledge that stationed itself either between humanity and Earth or between humanity and heaven, effectively removing whatever quality and value there was at the outset. Some people bound their hands to a futile labour that merely resulted in destruction, while others imprisoned their minds with the absurd shackles of dogma, itself an invention of a humanity that has gone astray. In Nature there is no such dogma, because here only constant flow, constant movement and hence constant reciprocity prevails.

Whether those leaders, who led the broad mass of humanity astray, involuntarily deceived themselves and therefore others, can never be properly established. In any case, the time is not far off when people will become free and learn to understand the *ur*-purpose of life through their own perceptions. The means is already at hand to free humanity from the accursed greed and material enslavement to which the vast majority fell victim and in which we became ensnared because we unconditionally believed everything. The wherewithal is a machine that has nothing whatsoever in common with the nature of contemporary machines. This machine is organised Nature. It is able to transform various substances in a form-building sense, and to ennoble all materials of whatever kind.

The products of this ennoblement are pure ethereal substances possessed of a higher quality, which animate and increase all things. The ensuing superabundance of everything will release humanity from bondage. In this way humanity, now a robber and a glutton, will slowly divest itself of material needs and become again what it once was, the crown of Creation. It is the final link in this organically interwoven structure, the organiser and representative of Creation on Earth. According to an old saying, originally everything was already here, but it all had to happen twice, precisely because humanity, born as a duality, is both good and bad, wise and unspeakably stupid.

As was the case in former paradise, humanity, by nature never a beast of burden, will be delivered from all labour with my machine. Such a state of unemployment has nothing to do with the present world affliction. Instead of work (in its present sense) a sense of service will evolve for which humanity was destined since the beginning of time; service to Nature. Such a service encompasses the organisation and dispensing of substances that have materialised into the world, which humanity must then fairly and evenly distribute. These concentrations of energy, increasing to the point of superfluity, will be equally active everywhere. They will thereby create an equality which, in our dim recollections, we recognise as true kinship and mutual compatibility.

Pseudo-communism will pass away and in its place the true communism will arise, of which Christ taught two thousand years ago. It could not come about earlier because at the time all knowledge of the laws of Nature was kept secret by the élite of society, and because the broad mass of the people were too immature fully to comprehend the true teachings. For this reason humanity had to travel the bitter road of suffering and devise a technology which, as everything else that humanity does for the first time, is one-sided, incomplete and hence of no real value. On the brink of suicide, which in all certainty today's technology will mete out to us if we continue to make use of it, humanity has rediscovered the means to save itself from oblivion.

This machine will reduce the air-pressure that presently presses down on the Earth. In consequence abundance will be produced automatically, because a rebirth will occur of those things which first had to die because they were products of humanity's first efforts and therefore were one-sided. Through the rebirth of spirit, humanity will become accustomed to revere and care for all life arising anew out of the Earth. We will emulate the bees, who are known to dispense and to give whenever they take.

Since we already know how to take and steal, we will have to learn how to give. In all our doings we will shun inorganic compounds (such as artificial fertiliser) and abandon the stupidity of obtaining energy and power by combusting the building-blocks of Nature. The organ-machine* will fulfil every wish that a person who desires to foster growth can conceive. It will, however, destroy everything that is untrue, speculative and deceitful. It was my duty to find this machine. The mighty task that now awaits is to find those people who understand how to organise everything in readiness for the coming upheaval of this era. But before these people have been found no human eye will ever see this machine in full operation, because I am fully aware of the awesome responsibility involved. It forces me to keep silent until I am convinced that this machine will serve only one purpose: Nature.

*Paved with extraordinary insights into Nature's deeper workings, Viktor Schauberger's long and arduous road through Nature's living laboratory towards the discovery of this machine will slowly be revealed in the pages that follow. – Ed.

Nature as Teacher

Nature as Teacher

From the combined texts from *Implosion* Magazine, No.7, No.19, No.111 and *TAU* Magazine, No.146 (June 1936).

Nature moves everything and creates this movement through various differences in temper-ature and potential. Where these intersect the primordial life-force is born. There are only a very few people who really know how to observe Nature and to notice these small and almost imperceptible differences in motion, temperature, tension and potential. With them it is possible to overcome physical weight, to switch off the universally recognised Law of Gravity and to demonstrate that the same apparently irrefutable Law of Conservation of Energy is a fallacy. If these two laws, so fundamental to contemporary science and technology, are invalidated, then the crowning law of the latter, in which resistance to motion increases by the square of the increase in the speed of motion, can no longer be upheld. It goes without saying that the collapse of these conformities with law automatically negates the law of heat-equivalence and the law of gravity.

Nobody with a scientific education believes that these are real possibilities and will consider it all to be utopian nonsense. It is precisely these differences, however, that enable them to overcome their body-weight and makes possible their spiritual and intellectual development, growth, and ability to reproduce. For this reason it is necessary to give a few examples of these decisive differences in motion, temperature and potential. At their point of intersection *il primo movere*, the animating impulse, is awakened in any material which is moved naturalesquely. The principal aspects of what is involved here must be approached with circumspection in order to realise the necessity of changing our way of thinking completely. Or, more accurately, of thinking one octave higher.

Not everyone will be successful in this endeavour. Although some people will be puzzled and begin to wonder, many will reject what follows. They will feel that their existence, their beliefs, their entrepreneurial or other opportunities and the very bases of their lives, are being threatened. Only those with exceptional intuitive abilities, and thus of an artistic turn of mind, will actually be able to grapple with this extremely difficult way of thinking. The initial propositions are an introduction to the examples and allegories that now follow. For instance, in order to get some idea of the importance of such an apparently insignificant difference in motion to overcome the dead point. This depends principally on the intensification of the *ur*-primordial and *ur*-propagative motive force.

Natural phenomena, still undisturbed by human hand, provide us with the clues to create a new kind of technology, for which a well-developed ability for observation is necessary. First we have to understand Nature if we wish to copy her systems of motion. As *Wildmeister* (Master of the Wilderness) in a remote forest reserve rarely visited by human beings, I was able to observe such events. Ultimately they led me to the discovery of implosion.

The Ödsee Rumbles
From *Implosion* Magazine, No.7, p.21 "Nature as Teacher"

Below the Ring in the Hetzau lie the Ödseen (Öd lakes). After a long period of hot weather they begin to rumble. This is the local expression for the thunder-like noise that wells up from the bottom of the Öd lakes, when water-spouts as high as a house rise out of the middle of them. I will describe the event as I witnessed it.

One hot summer's day I sat at the edge of the lake, debating whether I would go for a cooling swim. I had decided to do so when I noticed that the lake water had begun to swirl about in strange whorls. Trees, which had been carried down into the lake by avalanches, branches and all, freed themselves from being stuck in the sand and began to perform a spiral dance, which swept them faster and faster towards the centre of the lake. There they suddenly stood up on end and were dragged down into the depths by such a powerful suctional force that their bark was stripped off. This is similar to instances where people have been carried aloft by tornadoes fully clothed, only to be cast down to Earth again stark naked. Not a single tree ever reappeared from the depths of the lake.

Shortly thereafter the lake was calm again as if its hunger had been satisfied by the victims it had sucked into its maw. This, however, was merely the calm before the actual storm. Suddenly the lake-bottom began to bellow. All at once a water-spout as tall as a house shot up into the air from the middle of the lake. This turning, twisting, funnel-shaped, up-surge of water was accompanied by a thunderous roar. Then it collapsed. Waves struck the shore, which I had to leave as quickly as possible because the water-level in the lake suddenly began to rise uncannily. What I had just witnessed was the 'primordial' growth of water, the renewal of water in lakes without affluent streams. For the first time I began to understand what was happening, though not all was completely clear to me, and another experience was necessary to clarify this intriguing question.

The Fish-Eagle
From *Implosion* Magazine, No.7, p.21 "Nature as Teacher"

As a young apprentice forester I spent every free moment I could in a sleek rowing boat on a mountain lake near the forester's house. I caught fish and hunted ducks and other aquatic game.[23] My dearest wish was to shoot a mighty fish-eagle. This magnificent bird appeared every evening above the so-called salmon holes, circled in peculiar curves and then fell like a stone, only to soar upwards again with a large salmon in its talons.

What always puzzled me was how the fish-eagle managed to catch fish in the salmon-holes without actually diving into the water, because the salmon never came to the surface. It was a puzzle that took me a long time to solve. The only solution, I thought to myself, was to observe the predator and its victims very carefully. A steep outcrop of rock on top of which there was a tall spruce tree provided an ideal vantage point. Equipped with a good spyglass and well concealed by upper foliage, I sat in the top of this tree alternately watching the eagle and the salmon swimming in the hole, which lay at an angle below me. I was careful not to make the slightest movement.

Punctual almost to the minute, the mighty fish-eagle made its appearance and flew over the salmon holes just above the water, emitting a shrill cry and flapping its wings particularly vigorously. It was as though it wanted to tell its victims that it was now there and that they should prepare a suitable offering. The eagle then rose almost vertically in smaller and smaller spiralling circles, altering its course with an occasional flap of the wings. Then like a block of stone, with wings furled and talons tight against its body, it plummeted down onto one of the salmon-holes. Just above the water-surface it braked its fall with a flap of the wings and a large salmon was already wriggling in it claws. Heavily laden, it flew towards the forest in a wide curve and disappeared. Quite logically, my immediate attention was drawn to the behaviour of the fish-eagle. I totally forgot about the

---

[23] Schauberger wrote this and later hunting references in a time when hunting was not ecologically disapproved of in the way it is now. In his day it was still conceivable simultaneously to love Nature and to hunt. Species were not then so rare or threatened as now, and old rural traditions, now dead, still survived. Values have changed! – Ed.

behaviour of the apparently well-protected and largely motionless salmon in the deep holes.

The next time I determined to observe everything very closely indeed. The eagle arrived, made its presence known as before and soared aloft in the now familiar, increasingly tight spiral curves. The spectacle was so fascinating that I nearly fell over backwards, because my body was unconsciously following the movements of the fish below. I only just managed to catch myself in time. All the salmon were copying the upward-spiralling, swirling movements of the fish-eagle above them. As though joined together with a string, they followed each other towards the surface, describing narrower and narrower spiral turns in their circular ascent. The fish swimming in the middle were pressed so closely together that their fins stuck out of the water.

There it was again? The dark shadow of the eagle and the miniature maelstrom. The eagle had one of the largest salmon in its clutches and sailed off. As often as I was able, I watched this performance, which always followed the same sequence. Every time I was surprised at my own reactions while sitting on my perch high up in the air. As if hypnotised, I faithfully copied the movements of the fish like a transfixed child watching a swing or the grimaces of a clown. I then decided to shoot the fish-eagle. The next time it swooped down on the salmon-holes I fired at it with my shotgun. The eagle floated, wings outstretched, on the surface of the water and I quickly reached it in my boat. When I looked at it sitting there so helplessly, I decided then and there to catch the majestic bird alive and keep it in a large chicken coop.

It was not as easy as it looked, however. I drew closer and touched it with the tip of an oar from which splinters were already flying. Finally I got out my short leather jacket, buttoned it up carefully and dropped it over the eagle's head. Dripping with sweat and bleeding from many scratches, I returned home, where it took a combined effort to encage this exhausted eagle. After a few days it began to take food again. Fish and water-snakes were in ample supply and the wounded wing healed quickly. Eventually we became good friends and later on I let it go.

The Swing
From *Implosion* Magazine, No.111, p.24 "The Trout Motor"

One day a spoilt and stubborn child went for a walk in the garden with her over-anxious father. The bored child saw the long-forgotten swing and wanted to get on it. Hurrying after the running child, the anxious father sat her carefully on the seat and told his moody offspring how to sit on it properly. He then swung his little daughter carefully backwards and forwards, while she just sat there stiffly and unbendingly like some figure from a tragicomedy. After a short time both had enough of this unrewarding activity and side by side they continued their unhappy walk, if nothing more than to stimulate their appetites by the movement.

They had scarcely disappeared before a poor, wild little girl, entirely left to her own devices, who had watched the uninspired goings-on, stood in front of the forbidden and therefore much envied swing. After a quick look around that the coast was clear, the tomboy was instantly astride the mechanically well-built swing. She pulled her feet in and arched her body far back, swung forwards, shifted her weight, swung backwards and after a few swings stood up on the brown polished seat with legs wide apart and as she swung to and fro allowed her sunburnt hands to slide up and down the ropes. In so doing the point was soon reached where the ropes were parallel to the ground. She then changed her position again. She brought her feet together and spread her arms so as to rotate herself rhythmically about the longitudinal axis of her body, to the right on the foreswing and to the left on the backswing, all the while remembering to shift her weight forwards and backwards. In this way the swinging action turned into a round dance. The length of each swing became longer and longer and she swung higher and higher. A few moments later the ropes were almost vertical. Backwards again one last time and just before the point of culmination was reached, the child quickly flung herself in the opposite direction about the longitudinal axis,[24] thereby abruptly braking her body, which wanted to continue its original path about this axis. The effect of this purely intuitive manoeuvre was amazing. Her body hurtled forwards and, as occurs when a fast-sliding toboggan collides with an obstacle, was thrown in a high curve over the top rail of the swing, i.e. the earlier back and forth motion had become circular. In this faster and faster circular flight the backwards and forwards motion of the body become smaller and smaller and ultimately only a slight swaying back and forth could be observed. This took place at longer and longer intervals, finally resulting in a rhythmical looping motion, the difference between height and width of swing becoming almost indistinguishable, because the body now winds itself like a snake about the longitudinal axis.

Eventually the child grew tired of this activity which was so fascinating to the onlooker. A stiff jerk went through the child's body. The whole swing shook from the sudden braking recoil. The cycles became noticeably slower. Now the second critical moment arrived, namely the necessary damping of

---

[24] Here is some difficulty in describing this process accurately since there are no accompanying diagrams. Here the longitudinal axis would appear to be the one about which the swing rotates, namely the top rail from which it is suspended. – Ed

the momentum before the 'dead' point of culmination was reached. At the actual moment all that could be noticed was a twitching, somewhat angular movement of the body. Even the effect of this rhythm-destroying movement is astonishing. For a moment the seat of the swing hung motionless in the air almost a whole metre below the culmination point before swinging downwards once more. In the same instant the child returned to the seat and allowed the swing to move without influencing it further, until with shorter and shorter swings, alternately to the right and then to the left,[25] it eventually came to rest. The child gave the impression that she had come back to Earth from another world. Slowly brushing her hair back, she looked about rather nervously, tilted her head from one side to the other and listened with lowered eyebrows, thereby unconsciously intercepting all vital sound waves through the slightly inclined attitude of her head. A moment later she stood in front of the swing and remembered to bring it to a complete standstill. In a few leaps and bounds she reached the hedge and bending down, slipped through it like an eel. Apart from the slight swaying of the swing, nothing remained to betray the fact that this nature-attuned child had solved a problem over which countless astute thinkers had racked their brains in vain.

This concerns the curve or curved path with which it is possible to over-come the 'dead' point. This curve has been particularly well mastered by the millipede, which would hardly be able to walk at all were it ever to reflect on how and when to move its 1,000 feet at the right time. Even the slow snail would barely be able to move were it to ponder the problem of setting its caterpillar-like locomotive works in motion such that movement in a straight line would result from the sinuous movement of its underbelly. The same sort of movement also enables the pike, with a looping flick of its tail, to dart forwards like an arrow to seize its prey.

And now to a second example, which elucidates the pro-cess[26] in a some-what different light.

The Trout

From *Implosion* Magazine, No.19, "Our Motion is Wrong" & No.7, "Nature as Teacher"

I would like to explain, in a very unscientific way, how the old foresters, huntsmen and fishermen and their wives at the spinning-wheel, recounted old fairy-tales as they sought to explain the mysterious perseverance of the

---

[25] This would appear to confirm the interpretation that the observer was looking at the swing from the side and that the longitudinal axis of the swing is therefore centred along the top rail. – Ed.

[26] Sometimes Viktor separates prefixes with a hyphen through which the function of the prefix is accentuated – Ed

trout standing motionless in the axis of flowing springwater. It was rumoured that in such springs the souls of the departed were to be found. They were supposed to be slowly released from their decomposing earthly remains in the cool womb of Mother Earth. Gradually coming together in descending streams and migrating upstream along them, the souls became charged with a purifying ascending current upon reaching the spring. This current also draws up a little springwater in its wake, which then enables it to rise heavenwards. As a four-year old, I wanted to have a close look at this migration of souls and in the process fell into the ice-cold water. The maid-servant fished me out and administered stout blows, and in doing so she shook the water out of my lungs and stomach. She then took me into the kitchen and angrily sat me down on the kitchen sideboard, where my shocked mother took over. While my clothes were being changed, she delivered an unforgettable lecture.

"You silly boy! How dare you go to the water! The poor souls of the departed migrate through the water towards the mountains and are resurrected at the springs and carried up to Heaven by the ur-force of all life. They entrance you, pull you in, you drown and die, and then you have to go with them. When you are grown up, but not before, you can go to the water's edge. When you're older, and you have pressing problems and no longer know what to do or how to help yourself, go to the waters of mountain springs. There you will find me again, for there my soul will be. I will then give you motherly advice and help you when I am no longer on this Earth."

Thirty years later, as a young forester, I was going to lose my job due to the inexploitability of the timber in remote stands of the forest reserve. The forest was to be administered by a cheaper game-keeper instead. Standing near a crystal-clear mountain stream, I remembered my mother's words. Ruminatively I gazed into the fast-flowing water, which rose from a crack in the rocks a few hundred metres higher up. I murmured into it. Disappointed because my mother's spirit did not answer, I sought to cross over the metre-wide stream using my staff as a vaulting pole. I tried to find a secure purchase for the end of my staff on the smooth, rocky stream-bed. In the process I flushed a stationary trout from its lair, which fled upstream like lightning.

Two questions shot through my mind. Firstly, how do trout reach this high location, because a kilometre downstream the water plunges 60 metres (180ft) and is atomised into a veil of mist? Secondly, just how was it possible that there, right in the axis of the current, a large number of fish were able to stand so effortlessly motionless. They steered themselves with but slight movements of their tail-fins, both overcoming their own weight and the specific weight of the heavy water flowing against them? On closer inspection this became clearer. Was this movement of the trout the mysterious pointer given by my mother in relation to what she had once said? Was this the sign she promised me? Was it the souls migrating towards the mountains that drew the fish along in their wake? Or did an axial, biomagnetic force prevail which also prevented the downstream acceleration of heavy logs, and yet which, through mysterious counter-forces of suction, made their transport down the central axial possible? Suddenly I had a vision of a machine that would overcome gravity.

In this way I discovered the animalistic magnetism of earth, sap and blood, which enables the naturally flowing, planetarily inward-spiralling water-masses to maintain their steadiness of flow in variable gradients. This steadying force is rendered inoperative if the watercourse is regulated and straightened out. It is also extinguished if springwater is centrifuged in high-speed, steel pressure-turbines.

It was during the spawning season on a bright moonlit night in the early part of the year. I was sitting near a waterfall on the lookout for a dangerous fish poacher who used to throw bottles of unslaked lime into ponds immediately below the waterfalls. These ponds were always full of fish. The result of this was that whole streams were emptied of fish due to lime-explosions and the cauterising of the fishes' gills. What occurred that night happened so fast that there was hardly any time to register what actually took place. In the bright moonlight, which fell on the water at the right angle, every movement of the numerous fish could be discerned in the crystal-clear water.

Suddenly the fish scattered in all directions. The reason for their flight was the appearance of a particularly large trout which came up from below, made for the base of the waterfall and immediately began to swim around it. These movements, which seemed to describe the shape of an egg, were akin to those of the child mentioned previously, who swung on the swing in a naturalesque way. It seemed as though the trout was rocking itself to and fro in strongly pronounced looping movements. It was dancing a sort of reel in the swirling water. All at once it disappeared under the fall of water, which fell like liquid metal into the pond. The trout suddenly stood up on its tail and in the conically converging stream of water I perceived a wild movement like a spinning top, the cause of which was not immediately apparent. Having temporarily disappeared, the trout then re-emerged from this spinning movement and floated motionlessly upwards. Upon reaching the underside of the topmost curve of the waterfall it did a quick somersault in a high curve upstream (backwards like that of the child) and with a loud smack was thrown beyond the upper curvature. With a powerful flick of its tail-fins it disappeared.

My head full of thoughts, I filled my pipe and smoked it until it went out. All concern for the expected arrival of the poacher was forgotten. In deep thought I returned home. Later on I often saw how trout playfully surmounted high waterfalls, in this case 6 metres (19ft) high. How and why they were able to do this, however, I only discovered years later as a result of other observations which happened one after the other like a string of pearls. No scientist was ever able to explain this phenomenon to me.

The Ox

From *TAU* Magazine, No.146 (June 1939), p. 30 "The Ox & the Chamois"

Anyone, who has observed the appalling cruelty suffered by wretched draught-animals in the difficult and dangerous transport of logs in the mountains, will understand why I have concentrated my whole mind and thoughts on the possibility of bringing this, in any case worthless timber, down into the valley without the brains of oxen. My many suggestions for transporting timber by water were always rejected because this traditional system of transport usually causes such damage to streams that it is more economic to build forest roads and railways. They are excessively expensive to operate but nevertheless are cheaper than the subsequent damage which becomes evident in any stream-channel after a short time. Reference was always made to Archimedes' famous principle: heavy beech-logs do not float and so on scientific grounds alone it would be utopian to take into account such foolish suggestions for water-transport.

My father had transported hundreds of thousands of solid cubic metres of beech over large distances. He never allowed this work to be carried out by day but as a rule on moonlit nights. As my father always explained, water irradiated by the Sun becomes tired and lazy, rolls itself up and sleeps. Whereas, during the night and especially in moonlight, it becomes fresh and lively so that it is able to carry pine and beech logs heavier than water.

All my arguments for water transport were dismissed as ridiculous fantasies. Thus one day I found myself once again behind a heavily-overloaded log-sledge, which, under a torrent of abuse and cracks of the whip, the wretched team of oxen had to drag over the mountain,. The poor beasts careered downhill, spreading wide their four legs, eyes bulging with fear, because the teamster did not wish to apply brake-chains and spoil the good snow-track.

No friendly suggestions helped to alleviate the lot of the poor oxen under the yoke. Due to his paltry wages, the man was barely able to cover his expenses. What now happened I chalk up to a kindly fate with which wise Nature came to my assistance, without my having an inkling of its ramifications at the time. As a rule, the oxen, cringing with fear on the descent into the valley under withering whip-cracks, had to approach the next rise at the gallop, and on this occasion one of the oxen collapsed. Having become thoroughly entangled, all attempts to get this ox back on its feet failed. *"Shame you fell down! Now you'll just have to learn how to get up again!"* fumed the bemused peasant, and in a fury he began to strike the poor beast. The more I yelled at the angry man, the more he flailed at the animal. Finally the ox convulsed, dropping its foaming mouth onto the hard-packed snow. Out of its crushed eyes the last of its spirit glistened like tears.

The very next moment the man got out a halter with numerous silver clappers, which he passed over the standing oxen, sliding it over the humped back of the recumbent ox. Whereupon the blinded ox took such fright that with one heave it stood up on its four quivering legs again and in doing so tossed the peasant head over heels. The man took a long time to stand up again. Once this stalwart fellow had recovered his balance, he rubbed his nose backwards and forwards, right and left with the back of his hand. Quietly taking a sledge-hammer, he knocked away the retaining chains so that the logs tumbled thunderingly down the slope. *"Mr Forester, our agreement is at an end"*, he declared. Taking his oxen by the yoke-strap and grinning sardonically, he returned to the cabin to inform his colleagues of the happy end to his misbegotten contract.

A few hours later all the ox-teams left and I stood utterly alone in the forest, pondering on how I could get the already-sold timber down to the valley. Forced to act quickly, I had an improvised weir and sluice-gate erected, and without the slightest hesitation I had the logs thrown into the forest stream, which was beginning to rise owing to the meltwater. In the early morning the logs floated downstream without difficulty. The Sun had hardly blinked on the water before the logs sank and would not be budged from the spot. It was only in the late evening that the water came alive again, and in this way, avoiding certain times of day, I brought the logs, with only a few exceptions, down into the valley by night. Only in a deep ravine, where no ray of sunlight fell on the water, did the 'sinkers'[27] remain lying. No surge of water was able to stir these waterlogged timbers from where they lay. Curiously enough, in the following summer, however, on the occasion of a warm rain these stragglers turned up, whereupon I noticed that they no longer lay flat, but were floating *half upright*.

Accustomed to observe all natural phenomena very keenly, it quickly became clear to me that the Sun's rays are able to produce many effects undreamt of by our learned scientists. At any rate, the good Archimedes had overlooked many things in this regard so even in those days I had already lost faith in the ancient Greeks. It only became clear to me later on that old Isaac Newton, the deified Mayer[28] with his law of the conservation

[27] 'Sinkers': Colloquial term for logs heavier than water – Ed,
[28] This refers to the German physicist, Julius Robert von Mayer (1814-1878), who contributed to the formulation of the Law of Conservation of Energy. – Ed.

of energy, and the botanists with their law of heliotropism, had all erred. Once again, kind Fate came to my aid, and I will relate how this was manifested in a practical way.

Dancing Logs and Stones
From *Implosion* Magazine, No.7, "Dancing Stones" & *TAU* Magazine, No.146, "The Chamois"

On a particularly cold December morning after the above event had occurred, I was stalking a very powerful buck in the ravine where the logs always came to rest. It was a loner and seldom seen. I was just about to abandon the fruitless hunt when a small trickle of snow on the steeply-inclined walls caught my eye. Immediately afterwards the shrill warning call of the concealed chamois reached my ears, and at last I espied the much sought-after beast standing behind a dwarf pine. The situation was extremely tricky, because once shot, the chamois would inevitably plunge into the almost inaccessible ravine and in all probability smash its magnificent horns. In the end the hunting spirit got the upper hand. Under fire the good buck collapsed, slipped and fell head over heels over the edge. A moment later I heard it hit the ice-sheet at the bottom of the ravine far below with a dull thud.

"Damn!" I thought to myself, *"Now the horns will be broken and the beard will freeze and lose its beautiful symmetry!"*. It was hard to know what to do next. For hours I tried to find somewhere to climb down. Suddenly I slipped and tumbled into the ravine down the icy path of an avalanche. Using my staff as a brake, I landed luckily on a heap of snow. Overjoyed, I caught sight of the chamois, which fortunately had not fallen through the ice and into the water. A minute later I stood over it, and, admiring its horns and beard, I began to remove them. I lifted up its body and attached it securely to the carrying strap, taking its entrails over to an area of open water which lay a short way upstream, in order to repay the fish for my having disturbed their night's peace. Having slit open the entrails with a knife to release the gases, I watched them sink slowly towards the bottom until they finally lay on the sandy riverbed. At this point the water was several metres deep, crystal-clear and totally still. While washing my sweaty hands, I noticed many 'sinkers' about 4m (13ft) below. They were performing a remarkable dance. It was as though the logs lying in the absolutely still water had become magnetic. Now and then the butt of one log lifted, floated upwards slightly, laying itself a few feet over another log, only to recoil in the opposite direction the very next moment as if in fright. Chamois, horns and beard were forgotten! Hour upon hour passed. I was unable to drag myself away from this extraordinary spectacle. Gradually evening fell and with it

came an even keener chill. Suddenly a heavy log stood straight up, paused for a moment, and with a lurch shot out of the water, to be encircled immediately with an Elizabethan-style ruff of ice. Soon the other logs began to perform the same witches' dance. For a few minutes several logs, girdled with an ice necklace and projecting about 10cm (4 in.) above the surface, swayed to and fro in the water as if alive. Indeed, even more marvellous events were about totake place! For a long time I lay on the ice and observed the action at the bottom of the icy-cold stream. There! I couldn't believe my eyes! There was a strange activity amongst the variously-sized stones lying on the bed. A few were as large as a human head and began to come to life. For a long time they played the same game as the logs did earlier, first approaching each other only immediately to recoil. Flouting all the laws of gravity they drifted about, attracting and repelling each other. At first I thought that the stones might be electrically charged. I remembered the phenomenal luminescence[29] given off by such milky-coloured stones which leave yellowish-gold comet-tails behind them when rubbed together under water. This phenomenon is what had apparently given rise to the legendary Rhinegold in the 'Song of the Nibelungs'.

All of a sudden a large, head-sized stone began to gyrate slowly in circles like the trout at the waterfall before it begins to float upwards. It was egg-shaped. The very next moment the stone was at the surface right before my very eyes. It was quickly encircled by a visibly-growing collar of ice and this apparently-possessed, gently rocking stone, floated on the full-moonlit surface of the water, just as the heavy beech-logs had done earlier on. Soon a second, then a third, and still more stones, all making the same manoeuvres, rose to the surface of the water. The water soon looked like a cake bespeckled with raisins. Eventually almost all the milky-white stones that had been smoothed and rounded rose to the surface. The remaining rough, angular stones which had fallen in from the banks were left motionless on the bed of the stream.

At the time I took no account of the precise angle of the incident moonbeams, which make lunatics walk about on roofs. Naturally I had no idea that this involved a concentrative process, nor did I realise the significance of the oxygen-concentrating effects of the chill on this bitterly cold night of the full Moon. This cold concentration gives rise to the emission of expansive emanations, which lead to the creation of the 'original'[30] form of motion. It is this form of movement that overcomes gravity and raises the specifically

---

[29] This refers to triboluminescence, which is the energy, expressed as light, given off by electrons returning to their rest-orbits after having been excited due to the pressure arising when stones are rubbed together. This emission of energy is a nett gain to the overall energy-content of water. See *The Water Wizard*, "Fire under Water". – Ed.

[30] Here 'original' also means form- or structure-originating or form- or structure-bestowing motion – Ed.

heavier stones[31] to the surface of the water. As a huntsman I was well aware that female physical masses, charged with negative ions, become fiery when handled coolly. I also knew that in their hunger for reactive substances (semen or fertilisation) these female bodies, known to triumph over the male when they succumb, were able to overcome not only their own weight, but also that of their superimposed load. What I found hard to believe was that, apart from negating their own weight, these stones were also able to overpower the watery resistance to motion pressing down upon them.

Although at the time I was unaware that this process was the starting-point for the explanation of atomic transformation, it nevertheless made me decide to be even more observant in the future. As a result further events unfolded which, understandably, appear highly mystical to scientists, but which in reality are completely natural manifestations of the forces of Nature. Without them there would be no life and movement in this fascinating world. My main concern of the moment, however, was how to remove the chamois, which by this time had become firmly frozen in the ice, and how to climb out of the ravine in the middle of the night together with my buck. Luckily it was a glorious full Moon lit night. I turned my steps homeward over an ice-bridge that had formed. Many years later I learned that this natural phenomenon, which has still not been explained scientifically, was an everyday occurrence on the River Angara, the outflow from Lake Baikal, and enabled the farmers to cross this bridgeless river in winter.

In any event, I was fundamentally cured of whatever school knowledge I had left. On this day the relentless battle with established scientists began. I had unfortunately greatly underestimated these scientists, because I had no idea how dangerous is this species of human being. Without the faithful support of several women and the fighting comradeship of Werner Zimmermann I might perhaps have perished, as had many other discoverers before me. Thus this enquirer after truth, Zimmermann, also became part of my destiny. His *Tau* magazine brought me much joy and I am ever grateful to him for his constant encouragement. It greatly helped me to close the circle in order to show people previously unknown processes that Nature employs in her technology. Goethe was perhaps the only person who perceived these true processes of Nature. I quote:

All things into one are woven, each in each doth act and dwell
As cosmic forces, rising, falling, charging up this golden bell
With heaven-scented undulations, piercing Earth from powers sublime.
Harmonious all and all-resounding, fill they universe and time!

[31] This refers to the specific weight of the stones relative to the specific weight of the water, which equals 1 and is the base value for all definition of specific weight. In Viktor Schauberger's view it also refers to the difference between the weight of mattered imbued with life (specific weight) and dead matter (absolute weight). – Ed.

Amidst life's tides in raging motion I ebb and flood – I waft to and fro!
Birth and grave – eternal ocean, ever-moving, transient flow.
Such changing, vibrant animation – the very stuff of life – is mine,
Thus at the loom of time I sit and weave this living cloth divine.

How blossomingly I rejoice! All hail to the new!
All is born of water and upheld by water too!
Transpierced thus am I by beauty and by truth!
Oh great ocean, grant us thine eternal ruth!

Wouldst thou not send clouds, nor bounteous streams endow,
Nor perfect the currents, nor rivers here and there bestow,
Then where would mountains be, and what of plains and world?
For thou alone it is that keeps this freshest life unfurled.

It was a blinded ox, a shy chamois and finally it was Goethe, the prince of poets, who gave me eyes to see.

The Interchange of Substances in Nature

From *Implosion* Magazine, No.7, p. 6 "Nature as Teacher"

Most people have heard of the phenomenon of St Elmo's fire. It consists of bright flashes of light which emerge from the tips of treetops and give off a fiery glow. Why and how it occurs has so far not been explained. A similar phenomenon was shown to me, and again it was when I was hunting. As a newly-fledged *Wildmeister* I was given permission to shoot my first game-cock. Naturally it would have to be the largest in the 30,000 acre forest-*Eldorado* and I became more and more excited at the prospect. The old forest warden, who often gave me advice, had explained to me that there was a legendary mountain cock, an age-old campaigner, that lived in an almost inaccessible hidden valley. No imperial highness, no serene duke, no degenerate prince and no doddery old baron had ever managed to enter this valley. The entrance to this gorge had been sealed off by a rock-slide. Access was extremely dangerous and only possible with a climbing rope.

One afternoon I stood for a while in front of this sheer, insuperable barrier. To the right there was an almost vertical wall of rock and to the left a 100m (328ft) waterfall, with flat areas of rock-face on each side, as slippery as an eel. One slip and my earthly remains could be carried home in a handkerchief. I spent a long time throwing up the grappling hook before it finally became lodged at a height otherwise inaccessible from the ground. Whether it would hold or not in the crumbling, weather-beaten spur of rock was a matter of conjecture. After one or two attempts to tear it loose, the ascent began, step by step, and with trembling knees I finally reached the top. After a short rest I pressed on into the virginal high-mountain gorge in which no-one had set foot perhaps for decades.

The chamois I encountered during my climb eyed me warily, but did not see fit to give their piping, warning call. Apparently they were still unaware of the danger this huntsman represented, but they avoided this two-legged monster all the same. Indeed they were hardly bothered at all by their Wildmeister, which I found quite infuriating and because of this I struck out at a dwarf-pine bush from behind which an old chamois doe emerged with a flying leap. Instead of running away, however, it came closer out of curiosity and to have a better look at the odd creature that had disturbed its light feeding. Eventually I entered a small primeval forest, which spread over the few acres at the foot of the sheer north face of the crags and in which old spruce, large larches and other species of overstorey and understorey were growing. It was exactly what a proper primeval forest should be. After a brief search, I found the courtship display-ground and saw by the droppings that the old warrior perched on a gnarled branch every morning to sing his lovesong.

I carefully circled the roosting tree in order to find the right place to ambush the cock the following morning. By evening all I wanted to do was to bring the mountain-cock down with a not so harmonious and silent ball of shot. Meanwhile evening had come and darkness quickly fell. It was the night of the new Moon and was darker than I had ever experienced before. Half-sitting, I leant up against my bivouac tree, wrapped up in my weatherproof cloak and kept as still as a mouse, lest I frighten the cock away. A quite eerie silence descended. In the raven-black night it was almost impossible to see my outstretched hand. It was chilly and shivering, I huddled even further into the broad folds of my cloak.

From time to time I dozed off and lost all sense of time and place. It could well have been towards midnight that it all began. A small reddish flame flared up from the floor of the forest in front of me. At first I thought I had been careless when lighting my pipe and had set fire to the forest. However, I had been nowhere near the place where the fire started. It must be a will-o'-the-wisp, I thought, and continued to watch it. But when a fiery egg arose from the ground, I thought my eyes were playing tricks. With its narrower end pointing downwards, it stood quite immobile on a mound-like rise and gave off a pale yellow glow. I was already on my feet and stared at this uncanny spectacle, shaking with fright and cold. It grew larger and larger and eventually was about two metres high and a metre wide. It was a magnificent, but unearthly sight.

I wanted to run away as fast as I could, remembering the stories of the supernatural told at the spinning wheel on the winter nights of childhood. Having listened to them, the pretty or not so pretty farmer's daughters no longer dared to return home alone and were absolutely delighted whenever the young men accompanied them. It was not without ulterior motives that the latter recounted these horror stories, for in so doing they received their well-earned reward. Had these maidens not been willing, and this one cannot know, they might have impaled themselves on the horn of a different more awesome devil. Thus it was that the terrified girls generally chose the lesser of the two evils.

But where was I to run? All I could do was to take a firm grip on my shotgun and warily approach this glowing light. The closer I came, and I cannot deny it, the more my knees shook. And then I actually stood right next to this colourless light. It floated a few centimetres above a small mound covered with snow-white flowers. Carefully I held my steel-tipped staff in the flame. I smelled nothing and felt no heat. Becoming curious, I then held my hand in this glowing egg, but felt nothing, nor did I see the shadow of my hand. I then lifted the turf under the egg with the end of my staff. Nothing moved. Slowly I retreated to my tree, keeping my eyes fixed on the flame. Eventually it began to fade and go out. All at once the spook was over. It was replaced by a conspicuous warmth and gradually the dawn began to break.

When I saw the mountain-cock in the broad daylight, I lost all desire to shoot it. The old battler with his sparse tail-feathers and bedraggled plumage, looked thoroughly worn out. If stuffed it wouldn't have been much of a trophy and because of this I spared its life. A few days later I shot another cock, which to this day adorns my wall. Still pondering the events of the night before, I returned to the mound above which the night fire had flared. Delicate flowers of unparalleled beauty covered the whole mound, upon whose petals extra-large dewdrops stood. I had yet another experience, for when I touched them, they fell to the ground as though they had been struck by a blow.

I then investigated the ground underneath, lifting the sod with my staff. There was nothing to be seen and I was just about to cease my examination, when I felt a resistance. What I found was a large chamois horn. As I opened up the earth further, I found myself standing over a chamois gravesite filled with many, well-preserved horns. From old foresters I had already heard that the chamois, when they sense their impending death, seek out a common grave in out of the way places.

Before leaving, I covered the grave up again. It might have been an act of piety. Why I didn't dig up the rest of the almost undecomposed horns, I still don't know. But, I was enriched by an experience, however, which would later help to explain the secret of the transmutation of matter. As Fate had perhaps desired, I had been able to make another observation which was to be of crucial importance to the understanding of implosion.

The First Ecotechnical Practice

The First Ecotechnical Practice
From *Implosion* Magazine, No.7.

"In every case do the opposite to whatever technology does today. Then you will always be on the right track." V.S. (*Implosion* Magazine, No.36, p.3)

"Modern systems of motion are wrong! They are responsible for cancer and the unstoppable ecological ruin of humanity." V.S. (*Implosion* Magazine, No.51, p.23)

T he coat-of-arms of the Schauberger family consists of a broken tree-trunk entwined by a wild dog-rose. The family motto is *"Fidus in Silvis Silentibus"* (Be faithful to the silent forests).

Once upon a time there was a castle on the Schauberg. Betrayed, the last sons of this robber-baron were imprisoned by the Archbishop of Passau. The eldest was beheaded and the youngest pardoned and sent into exile. This last remaining Schauberger was banished to the primeval forests surrounding the Dreisesselberg. He settled on the shores of Lake Plockenstein and led the life of a forest bailiff in charge of law and order in the area under his jurisdiction. It was from him that the Schaubergers are descended, who uninterruptedly dwelt in this forest isolation as forest wardens and game-keepers, fishermen and foresters for almost a 1,000 years.

The wild dog-rose that wound itself about the broken trunk, from which it drew its life-force, symbolically depicted the more beautiful form of new life, which supports itself by entwining about the dead life-form. No doubt those who looked down from the mountain (schau = look; berg = mountain) were able to perceive the origins and causes of this eternal becoming and passing away, and with its blossom also inherited the gift of seeing into the most secret of Nature's processes.

Although they served under the bishop's crook for 1,000 years, they had just as little time for religious doctrine as they had for scientific facts. They relied on their own eyes and their inborn, intuitive abilities. Above all they knew of the inner healing power of water and understood how to achieve a striking increase in the yield of adjacent fields and pastures with the special shape of their irrigation channels, which however, were only operated at night. Their principal preoccupation, however, was directed towards the conservation of the forest and wild game.

From time to time floodwaters uncovered the very special bank structures of the mountain streams, which in these forest districts were used to raft timber. These constructions forced the water to rotate to right and left in serpentine, spiral curves. The concept of the 'cycloid-spiral space-curve' was naturally unknown to them. Strangely however, they used it ably enough in the construction of so-called timber or water chutes. Through the rhythmical alternation of these flow-directing devices both water and logs were endowed with such grace, that in certain sections of the flumes they even flowed uphill, flouting the laws of gravity.

We only need to cast our minds back to the symbolic snake of Asclepius or the biblical 'Tree of Knowledge'. Again and again we come across the coiling of a snake about a staff, a rod or a tree as the symbol of the particular type of movement associated with, and conducive to, cognitive power.

From earliest youth, all I wanted to do was to become a forester or a hunter, as my father, grandfather, great grandfather and great, great grandfather had been before me. My father wanted to force me to follow the higher career of a public servant. My mother on the other hand secretly supported me and always urged, "*Stay in the calling of your forebears, who in their day honoured the family motto, 'Be faithful to the silent forests'*". When I was introduced to the forest reserve by an old forester, my dream began to turn into reality in a way that I could not have imagined better.

After the end of the First World War I was able to take over the management of a hunting reserve because an old forester had been pensioned off. It was an extremely remote hunting estate, and because of this it was most beautiful. Although I didn't see eye to eye with my superior – a forestry commissioner who treated me very condescendingly – I quickly gained the confidence of Prince Adolf zu Schaumburg-Lippe, to whom the hunting estate belonged. He married shortly thereafter and I won the appreciation of the beautiful young Princess when, on her birthday she shot her first buck, a mighty twelve-pointer, which I had summoned for her with a conch-shell in true hunting tradition. One day while out stalking, the Princess confided to me that her husband was worried about losing his estate due to the changed circumstances of the post-war period. The Prince had had to forfeit much of his capital and could only keep his estate if it could be made to return a profit.

A long time before this happened, I had submitted my plan to the Estate Administration for reducing the cost of transporting timber by 90%. However, it had been rejected by my superiors as unfeasible, for they believed that in accordance with Archimedes' Law it would be impossible to float the heavy beech and oak logs, which were heavier than water. Coming from a middle-class background, the Princess had a more open mind. She had heard about my plan for a log-flume and intuitively felt that it might possibly be the saving of the hunting estate. She had also heard of the timber merchant's attempt to bribe me to reveal my plan to them. Quite out of the blue she asked me, *"How much did the timber merchant offer you?" "Three times my annual salary,"* I answered. Suddenly she stood still and asked me directly, *"What is your estimate of the savings in costs that could be achieved with your log-flume? Today it costs about 12 schillings per solid cubic metre to bring the logs down to the sawmill. 30,000 cubic metres are felled every year. This amounts to 360,000 schillings per annum. What would it cost if your installation were to be built?"* I replied, *"One schilling per cubic metre, including the cost of amortising the installation"*.

The Prince's annual contribution towards the upkeep of the hunting estate amounted to 80,000 schillings, and the Princess had quickly calculated the cost-savings, and with them the saving of the property. "Great!" she declared, *"It will be done!"*. My only stipulation, however, was that I was to have total freedom to act and negotiate the construction of the log-flume, for which I would also assume full responsibility. The Princess kept her word. I had no idea of the risk I had taken on. However, luck never failed me even in the trickiest situations. At the right moment I was able to make an observation, the far-reaching effects of which I was able to appreciate in their entirety only years later.

About four months later the flume was finished. The large logs were stacked in readiness. One day I made a preliminary test. An average-sized log was introduced into the mouth of the flume. It floated down about 100m and suddenly stuck on the bottom. The oncoming water backed up and the flume overflowed. I saw the faces full of scorn and malicious glee. I perceived the implications of this failure at once and was totally aghast. I had the stranded log removed. My diagnosis was that there was too little water and that the gradient was too steep. I was at a complete loss and sent all my workers home in order to ponder the problem in peace and quiet. The curves were properly arranged, of that there was no doubt. What was to blame for this malfunction? Those were my thoughts as I slowly walked alongside the flume until I reached the weir further down, where the logs were snigged and sorted and to which another flume was connected. The weir was full. I sat down on a projecting rock in the warm sunlight and looked down into the water.

Through my leather breeches I suddenly felt something scrabbling underneath me. I sprang up and saw a snake that had been coiled up on the spot where I had chosen to sit. I flung it away from me and it flew into the water, whereupon it swam straight towards the bank and tried to get out. It was unable to do so, however, because the rock face was too steep. In its search for a way out it cast about hither and thither and eventually swam right across the dam. I watched it and was suddenly struck by the thought: how can the snake swim as fast as an arrow without any fins? I took out my spy-glass and observed the peculiar looping movement of the snake's body under the crystal-clear water. A moment later it reached the far side and was gone.

For quite a while I stood there as though turned to stone. In my mind's eye I reviewed the snake's every movement, which had twisted about in such a curious fashion under the water. It consisted of a wave-like combination of vertical and horizontal curves. In a flash I understood the process. It was the snake that had showed me *double spiral motion*.

An hour later I had reached my workers in the hut just as they were cooking their meal. I gave them specific instructions. *"Finish your meal as quickly as you can! Three men are to go to the sawmill. Ask the manager for a wagon and bring about 300 larch slats to the upper weir!"* The men looked at me askance. The foreman, a Tyrolean, interjected, *"What do you want to do with the slats?"*. I cut him short and told him just to do what I had ordered. I took the master-builder with me and we went up to the induction weir. I told him, *"You will receive double your wages, and together with your men you will nail the slats to the walls of the flume in the way that I will show you, even if you have to work through the night by torch-light. I will be there myself all the while."* The master-builder shrugged his shoulders and nodded. A few hours later the wagon arrived, and also the man who had been sent to the forester's house to fetch the nails. The whole night was filled with the sound of hammers. I carefully inspected the counter-curves in the bends of the flume, which were to force the water to twist in the flume in the same way as the snake in the weir.

Towards midnight I returned home. There was a memorandum from the inspecting Chief Forestry Commissioner, to the effect that tomorrow at ten in the morning the Prince, Princess and several experts and specialists would be present to observe the trial run. I slowly laid the note aside and pondered what to do next. At eight in the morning the men could have finished their work, if they worked right through the night. I took my staff and gun and reached the site in an hour. While still some distance off I heard the sound of hammers and saw the glow of the forest workers' bicycle lamps, used for riding to work in the early hours of the morning. *"When will you be finished?"*, I called out to them from some distance. The master-builder estimated that it could be done by nine in the morning. I promised everyone that they would all receive three times their hourly rate of pay if they could finish by eight. They did it by 7.30.

I then said to them, *"Go and have your breakfast now, because I'll need you at the induction sluice at 9.30. There is still an hour's work to be done, and then you'll all have a day off on double pay."* The master-builder suggested that I also rest awhile, because I wouldn't be able to hold out much longer. I declined all further conversation with a wave of my hand and went over to the induction sluice. There I waited until my people returned, and shortly thereafter the Prince and Princess arrived together with my bitter adversaries, the experts and specialists.

I greeted the royal couple and the chief forestry commissioner. I never even gave the others a glance. The Princess looked at me with a worried air and the old rafting-master lolled against a post with a sardonic grin. I ordered the sluice-gate to be opened. Behind me my workers prodded several slender logs into the water, shoving a 90cm diameter log unobtrusively aside. "*No! No!*" the old rafting master suddenly yelled out, "*Put the heavy one in the water too!*". I gave a quick nod and slowly the log floated towards me, barely sticking out of the water. Soon it reached the flume intake and dammed the water, causing the water-level in the weir slowly to rise. Nobody uttered a word. Everyone was staring at the log as it rose up with the water. The very next instant the flume would have to overflow.

All at once a gurgling sound erupted; the heavy log swung slightly to the right and to the left, and twisting like a snake, it then sped off down the flume, head high, like an arrow. Having described an elegant curve, a few seconds later it was out of sight. Everybody stood looking after it. Shaking his head, the old rafting-master spat into the flume in a high, arching curve, fished his tobacco plug from his toothless mouth with his forefinger, threw it into the water and grunted, "*Goddammit, it actually works!*". Not having understood, the Prince enquired what the rafting-master had said. I stammered that it was only a quotation from Götz of Berlichingen.[32] "*Yes indeed, it certainly was!*" agreed the chief forestry commissioner quickly. Grabbing my staff and gun, I quickly took my leave and climbed up into the forest and was away. Once out of sight I sat down on a rock with gun and staff between my knees and mopped the cold sweat from my brow. "*Once only and never again!*" I thought to myself. Had I not sat on the snake at that crucial moment, my life would now not have been worth living.

A few days later the decree arrived making me the director of the large hunting and forest reserve. This was soon followed with visits by experts and other specialists from all over the world – even government ministers made their appearance. Shortly thereafter I was summoned to the Ministry of Agriculture in Vienna. It was then that the battle really started, which has continued to this very day. It was often very ugly and dirty. In any event, from the day that I promised the Princess that I would build the log-flume, I was never left in peace again.

Galileo Galilei discovered the rotation of the Earth about the Sun, to the great agitation of the scientific and religious worlds. The Pope threatened the discoverer with excommunication. Bishops and other high dignitaries of

---

[32] This is a euphemism to cover the use of four-letter words. – Ed.

the Church laid heavy charges against him, and arraigned him before the courts. Finally he was made to retract and forswear by his colleagues. *"Whatever you say, it still rotates!"* were supposed to have been his last words before he died.

I never paid any attention to such as Galilei, Archimedes, Robert Mayer, Isaac Newton, Pythagoras or any other formulator of energetic laws. I simply observed logs or other floating bodies, which, although heavier than water, under certain conditions swam like fish. I also observed those places where these same bodies sank to the bottom and stayed there, even where powerful tractive forces were active in water flowing as straight as an arrow.

It was the curious movements of the swimming snake that first demonstrated the double spiral movement to me and helped me to achieve major successes. Hundreds of thousands of the heaviest logs followed the first one, as if there was no difference in their specific weight.[33] By this time it was quite clear to me why my forebears had organised the rafting streams in such a way. They had formulated the rule of thumb that the timber must be floated and the water guided in the same way that a boar urinates.

In fact the completion of the whole timber flotation system was almost prevented. One day the chief forestry commissioner, who, absolutely determined to exert his authority, appeared on the site in company with the political director of the Forestry Administration. He forbade any further work, thereby placing the completion of the whole, well-conceived project at risk. I was forced to explain my whole concept and plan to a large investigating commission. The commission was made up of senior forestry officials under the chairmanship of the chief officer of local government, all of whom were antagonistic towards me.

These gentlemen maintained that an 18m high dam wall I had designed and built would not be able to withstand the pressure of water. It would collapse when the mass of water exerted pressure on this far too flimsy dam wall. The danger therefore arose that a disaster of unimaginable proportions would befall the villages lying below it. I watched and listened to the uncompromising attitude of the investigating commissioners. I was aware of the purpose of the investigation and had duly taken the appropriate precautionary measures. They wanted to bring about my downfall and put an end to the entire log-flume installation. Because of this I never responded, even to the chief officer himself, but quickly descended the steps leading to

---

[33] "In the 2km long, constant, half-egg (point downwards) profiled, wooden log flume at Neuberg (Steiermark), equipped with wooden, rib-like, current guides, the temperature and velocity of the water were measured. In the morning, with a water temperature of between +9°C and +10°C it took a log about 29 minutes to cover the distance. At midday, with a temperature of between +13°C and +15°C, however, it took 40 minutes to cover the same distance under otherwise identical conditions." – V.S. (*Implosion* Magazine, No.88, p.5)

the dam. I took a shotgun with me and fired both barrels upstream. This was the signal for the huntsman positioned at the fully-charged upper weir, who was loyal to me, to open the large sluice-gate and let the water out. Because my only response to the chief officer's questions was to fire two shots, the commission apparently thought that I had gone mad. They demanded that I put the gun down immediately and to come up to the platform on which the wildly gesticulating members of the commission were standing.

Suddenly a roaring noise could be heard getting closer and closer. I pointed upstream and around the bend a 6m high churning wall of muddy water was approaching, in which a motley assortment of logs, tree-trunks and tufts of grass were floating. *"For God's sake, get up here at once!"*, yelled the chief officer. The commissioners waved their arms about madly, shouting at each other. From where I stood, they all looked like lunatics. I glanced sideways at these agitated people, and standing at the centre of the dam wall, leant over it, apparently extremely interested in the height and supposed instability of the wall onto which the flood of water was about to impact.

At least, that is what the commissioners thought, who were standing above me with bated breath. The water did not do this. On the contrary, it hit the wall very gently and, turning around, thrust itself with enormous force against the wild onslaught of the floodwater. In the process the accompanying logs were stood up almost vertically and leapt out of the maelstrom like fish (fig. 2). The dam's million cubic metre storage capacity was quickly filled and the dam-wall held. Nobody dared to come down to where I was standing. The huntsman who had let the water go at my signal then appeared and, staring at the motionless figures of the commission, jumped down the steps and asked, *"Well, did it work?"*. I nodded briefly, took out the empty cartridges, reloaded the gun and slowly climbed back up the steps.

"Well", declared the chief officer, *"More luck than good sense!"*. I looked at him from the side, allowing my gaze to follow the contours of his body and remarked, *"Sir, I believe that the lack of good sense lies on the other side"*. Turning quickly, I retraced my steps. The commissioners followed, talking rapidly amongst themselves, and when they had finished their discussion, the chief officer called after me to say that experts would be on their way from the head office in Vienna in order to examine the dam wall, and to determine its stability by exact static calculation, for although the wall actually held, it was not strong enough. The experts and commissioners came again a week later. They duly carried out their investigation, made their calculations and came to the conclusion that because of its ingenious construction, it had a safety factor of twelve – it was twelve times stronger than it need be.

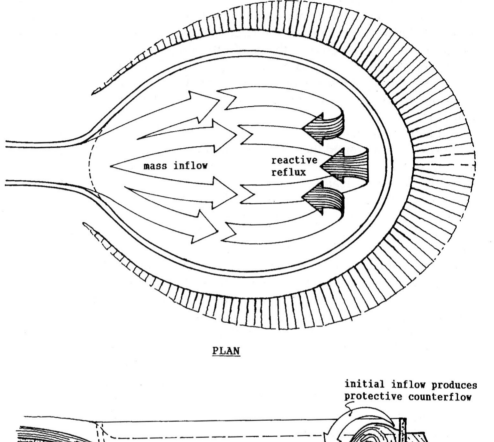

PLAN

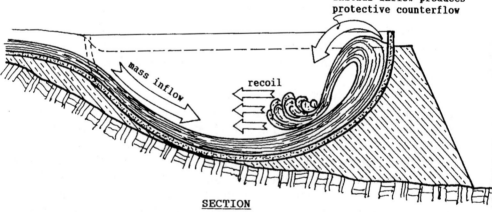

initial inflow produces
protective counterflow

mass inflow

recoil

SECTION

Fig. 2: The dynamics of the egg-shaped holding basin.[34]

[34] This figure has been reproduced from the editor's own book, *Living Energies*, for greater clarification. – Ed.

I demonstrated the filling of the dam again for the benefit of these men who, despite all their mathematical calculations, were still unable to explain everything. I drew their attention to the behaviour of the initial surge of water against the dam wall and its resultant recoil, whose counter-pressure neutralised the approaching main pressure. I had obtained the profile of the wall not from a university, nor from technical textbooks, but from a chicken's egg. Due to its egg shape, the circulating backwash off the 18m (59ft) high dam-wall, arising from the initial inflow, defended the wall against the otherwise-crushing pressure of the oncoming water.

Experience is Only Gained the Hard Way

From *Implosion* Magazine, No.7, p.21 – "Nature as Teacher"

It was a few years later, and another flume-weir had been built. During particularly heavy rain, a sluice-gate I had designed with a particular profile was being tested for the first time and the volume of flow carefully measured. A double-spiral pipe was attached to this special gate, in which the water was made to twist around itself like a stream of urine. With this I was able to increase the flow-velocity as the head of pressure reduced. This was made possible through the generation of a spiral suction-vortex, at the centre of which a white, shimmering reflux funnel[35] developed, for which there is a special cause. In this refluent canal [reflux funnel], clearly visible in almost all vortices in water, qualigenic formations[36] (ions) of an etheric and energetic nature move upstream. If these refluent substances encounter a cooler external temperature lying above them, then what I experienced at the time results.

It was raining in sheets. I lay on the float, looking down into the powerful vortex. As though falling from a gutter, the rainwater flowed from the upturned brim of my hat and into the hole. According to the conventional theory of gravity it ought to have fallen into the vortex. However, it did not. As it fell it widened out into a cone, broader at the base, and in this way a hat-shaped funnel formed above the lower vortex, the latter narrowing towards the bottom (fig. 3). I observed this extraordinary phenomenon very intently. But not for long, because suddenly an ice-cold jet of water struck me right in the face. The reason was soon evident. If the water-forming ethericities flowing backwards up the refluent canal encounter an air

---

[35] 'Reflux funnel': a channel or cavity in water in which the energies move in the opposite direction to the flow of the water (as seen when flowing down a drain), and into which water can be sucked up . – Ed.

[36] 'Qualigenic formations': agglomerations of quality-generating substances of an ethereal nature. – Ed.

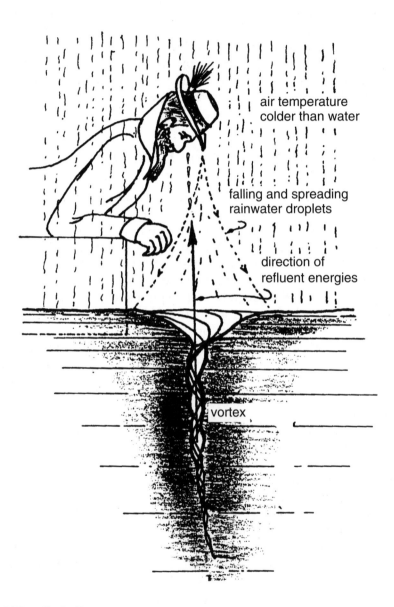

Fig. 3: The reflux jet.[37]

---

[37] This illustration has been produced by the editor for better comprehension of the phenomenon. – Ed.

temperature lower than that of the water, then the substances in which negative ions are concentrated radiate expansively and prevent the entry of the water falling in the opposite direction. In this situation a reactive upward suction develops.

The Winding Way to Wisdom
From *Implosion* Magazine, No.27.

In reputedly erudite circles I am considered mad. I was supposed to have been officially certified as insane and sent to the lunatic asylum in Mauer-Öhling at a relatively young age because of a dare-devil stunt I perpetrated on certain esteemed members of the Institute of Hydraulics. I was spared this fate however, because something actually worked that no one had thought remotely possible and undoubtedly baffled many intelligent minds.[38] The dam-wall I designed and built which, according to calculation ought to have burst, actually held .

In my log-flumes heavy timbers floated which, according to Archimedes' principle, ought to have sunk to the bottom, or at least to have become stranded on the sides. Despite this the logs in my flume floated down the centre. Flouting all the laws of centrifugal force, they described elegant inner curves when they ought to have followed the outer curve of the flume-walls. This upset the whole apple-cart of the supposedly sacrosanct laws of hydraulics.

Becoming somewhat over-emboldened as a result, I declared both verbally and in published articles that science was nothing more than a fraternity dangerous to the general public. It not only bred lunatics but also saw to it that they were well paid. Shortly thereafter, under the guise of a medical examination concerning my war pension, I was cleverly inveigled into a Viennese lunatic asylum. I only noticed when it was already too late. There I found myself among sixty genuine mental cases. There too were others who, in 1939, still believed that they could extricate the good old Austria of yore from the clutches of the Nazis, who had occupied it in a surprise invasion the year before. It was not exactly a pleasant experience.

In any event, it was a very strange feeling when, with an unimpeachable certificate in my pocket stating that I was fully possessed of my senses, I

---

[38] This refers to the egg-shaped holding basin, described previously, whose construction contravened all the law of hydraulics. He was told by experts that it would collapse when filled with water. To prove them wrong he stood on the wall itself just before the water was let in, declaring that if it failed then they would see his theories were wrong and that the world would be rid of another fool. – Ed.

was able to buy my ticket and travel homewards again. There everyone eyed me apprehensively. They knew why I had been taken to the place whence only a few rarely if ever returned. They wondered whether I still dared to champion the ideas which had got me into difficulty; whether I would still assert that it was not just any kind of 'poor soul' that went up to Heaven when we die. Rather, it was that which, as a metaphysical reversionary[39] influence, we have to thank for autarchical (original) motion.[40] In order to facilitate an understanding of this hitherto nameless and all-important entity, the following is written down in the way I first perceived and experienced it as a forester and hunter. Other equally important reversionary influences will be considered. These are important in their role as the actual metaphysical products of growth and transformation.

I now know what fate actually intended for me and why it caused me to experience all that I have. Because of this I came to understand a world in which no physical law of gravity exists. Owing to the levitational forces prevailing in this world of ours, that special quasi-material, quasi-immaterial substance has to float. As atomic fallout or reverting matter, it is actually heavier than its own medium. At higher altitudes this reverting matter no longer exists, for up there it is consumed by another force which, after such refreshment, becomes what I sought for decades as the 'specific and reactive counterforce to ever-active gravity'. Through this I came to understand *why* we actually have to live. I also became aware of the *purpose* of our mysterious journey through life. It is an interesting question that I feel reasonably equipped to answer, because my approach to the problem is different from other people's. For this reason some think me crazy.

One day I said to myself, "*If there is a force of gravity, then indubitably there has to be a force of levitation*". Since then I have been obsessed with this mysterious prior force, remembering a familiar remonstration from childhood, "*If you hadn't climbed up, you wouldn't have fallen down!*" If it were possible to discover this pre-active, antecedent force, I mused further, one would only have to switch it off at the right moment. After a brief counteraction by the familiar force of gravity, it would have to be switched on again. At that very instant the most marvellous of marvels, perpetual motion, would be accomplished and ready for use.

Long years spent observing what we commonly call 'growth' led to the realisation that Nature has operated in this levity-gravity alternating way since time began. Nature directs the warm rays of the Sun onto whatever is

---

[39] 'Reversionary': Describes the process of either reverting from the physical to the purely energetic or spiritual, or *vice versa*. – Ed.
[40] 'Autarchical': In this sense also self-organising, form- or structure-originating or form- or structure-bestowing motion – Ed

resurrected from the Earth, whereupon all that is of too inferior a quality *solidifies* under the influence of falling and concentrating light and heat. These solidified residues then manifest themselves as a net gain, which we need as food or as the driving force for our own bodies. From the residues thus bequeathed to us, we have to extract, root out or otherwise break down and ingest our own life-force in order effortlessly to overcome our own body-weight.

True growth, expressed as ripening, mating, reproducing, ageing and growing colder, extinguishing the life-force and relapsing into the great abdomen of the Earth, people consider a matter of course. Up to now, however, no one has ever taken the trouble to investigate what our dear old Earth then does in order to remain eternally young, and to be so burgeoning every spring.

I was then able to ponder about the entity to which the reactive rays of the Sun could do no harm, because it was the 'more sublime'. Goethe called it the 'Eternally Female'. This was of particular interest to me because he also called it the 'All-Uplifting'. It builds up everything around us that arches over our heads and into which day by day the Sun smiles. Perhaps it does so because physicists and technologists are quite incapable of conceiving how our Lord ever managed to create the ever new and blossoming splendour of springtime. Having ripened, whatever is too heavy wilts back into the womb of Mother Earth, only to smile at us anew in exalted form the following spring. It augments and ameliorates all that arches above us, into which the Sun roars with laughter.

Up there etheric forms and even higher entities abound. I just about managed to learn this while still at school. These higher embodiments of evolution not only rise upwards autarchically, but also draw a column of mercury up with them, which is otherwise known to sink when air-masses become heavy. Thus it was that this peculiar suction-force, which comes into being through a special form of motion called 'cycloid spiral space-curve motion', gave me much food for thought.

Because of all this I gradually began to be filled with doubts. I could no longer fully believe in everything I was taught at school and in church. I therefore became accustomed to regard each and every small event, even if apparently quite insignificant, as a perhaps not entirely accidental clue. Accordingly, each venture into the hunting reserve was no longer just a trip for its own sake, or undertaken to catch a poacher red-handed. On the contrary, every excursion became a voyage of discovery into almost hermetically-sealed and remote hunting areas, where hardly anyone had set foot before me. Much of what I saw, heard or otherwise perceived I did not immediately understand. Nevertheless, over the course of the years I was able to put the final polish on a completely new experiential paradigm.

One day, and this will be elaborated more closely later, I had two experiences in quick succession. I never forgot them and I can still see them very clearly in my mind. They set me many more puzzles to solve. I was on the point of jumping over a high mountain stream with the aid of my staff. While seeking a secure hold on the smooth and slippery rock bed for its tip, I scared a large trout from its lair where it is able to feed and rest without effort. As if no law of gravity existed and as though shot from a bow, the trout darted upstream like lightning. Two questions flashed into my mind just as quickly as the trout sped upstream.

One: How did this trout actually get to this spot (later I saw dozens of them in the same stream) which was cut off by a waterfall about 100 metres high roughly a kilometre downstream?

Two: What forces enabled the trout, not only to overcome its own body-weight so effortlessly and quickly, but also to overcome the weight of the water flowing against it?

Many years later I finally understood the cause of this phenomenon. It is the 'cycloid-spiral space-curve'. In conjunction with reactive differences in temperature it achieves the incredible; the overcoming of all weight through the creation of a reactive counterweight, generally known as the 'specific weight'. It is necessary to understand this concept as a condition of *densation* pertaining to higher, more etherealised and more energetic formative entities, already in a metaphysical state of being.

Having realised this, however, a great many more observations would have to be made of all the interactions before the true causes of this levitation could be appreciated. It was far more difficult, however, to construct the appropriate apparatus and to determine the proper alloys with which to build it. This had first to be done in order to establish the innumerable preconditions that were eventually crowned with success.

My obsession with this levitational force, would have remained fruitless had I not come to understand the whole reason for the creation of the human body. Through continual study of the processes taking place after death, I also realised that present religious ideas and world-views were either the perpetration of fraud on a grand scale, or self-deception to which we had become accustomed over the course of millennia.

I say this because anyone who is convinced that metaphysical growth is possible without continual decomposition and restructuring of earthly remains through 'cycloid-spiral space-curve motion', will never get any closer to the truth; to the nature of the genuine spiritually awakening force brought about by higher reversionary influences. Thus it was that one day I came to understand the 'hereafter'. It is the great vault arching above us that a famous poet once called the 'Living Breath of God', which is no empty phrase. Its significance will only become clear to those who know how to free the immense energies, material and immaterial, that are contained in every drop of water and in the smallest current of air.

The words 'latent state' are inadequate to describe the essential nature of this metaphysical energy-form. Our Indo-Germanic ancestors associated it with lurking demons and benign spirits that wait for just the right impulsive moment to re-enter their lookalike descendants and impel them into some form of activity. Anyone who cannot receive these higher, refluent inspirations has no intuition, is abandoned by all benign spirits, and falls victim to these demoniac forces. Such a person is then simply a ponderer, an *after*-thinker, and not a *fore*-seer or someone who intuits far into the future. Such a person will never understand what forces come into being when the paths of the 'disappearing' and the 'reappearing' intersect.

One breath of air, produced through the interaction of active and reactive forms of temperature, can bowl over an elephant that has merely become slightly overheated. This is because small changes in its metabolic processes interfere with the elephant's build up of its tremendous physical strength. This not only stems from the calories contained in active energy-concentrates, but in particular from reactive fields of *qualigen*. Moreover, the forces in question endow not only the elephant, but all that crawls and flies, with the original life-force. This includes the Earth itself which feeds all that it has created.

This experiential mystery ride is therefore only a means to an end, its purpose to lay the foundations for the real build-up of energetic matter. The body's forces are on short-term loan and required for mobility. They have to assemble, transform and accumulate all that is required for the build-up of higher quality substances from the remnants of this earlier organisation of raw materials.

The fundamental proposition that nothing can be created out of nothing was formulated by science. However, it failed to recognise the form of motion through which a densified product of concentration becomes manifest through the agency of the relatively highest-grade expansive motion. From every pore of this cycloidally-moving Earth a new entity ascends to take its place. It unfolds itself in the ante-chamber of the 'hereafter'. With the stimulating assistance of oligo-dynamic[41], decay-furthering and catalytic recombination-enhancing influences, this negatively charged, dynamite-like formative substance consumes the fertilising precipitates of solar energies. These solar energies must be further dissociated and dispersed by wind and other physical and mechanical dynamic impulses, becoming over-cooled in the process, in order to become the sustenance of growth and activity.

---

[41] 'Oligo-dynamic': Refers to energetic processes or physical products produced or triggered by small or subtle forces, which despite their apparent magnitude may also be extremely powerful. – Ed.

It would take too long to describe here the marvellous processes and roundabout routes through which the appropriate, nutrient-seeking entity must pass before attaining the condition generally referred to as a 'latent energy state'. Here is meant a potential state of the specifically highest density. Compressed into the smallest space and in the shortest intervals of time, *breathing* embryos of qualigen await these nutrient-seekers in order, unfolding anew, to grow upwards into even higher planes of evolution. In all activities which further development and converge progressively and increasingly intensely towards the Infinite, ballast must be discarded in order to give the 'primary' fresh upward impetus. This lower-grade (secondary) product of synthesis solidifies and serves as the physical nourishment for everything destined to vegetate. These exalted, vitalising forces are secreted in every particle of water and every breath of air and are activated by means of cycloidally-oscillating organs. Through the cycloid circulation of the blood, the dissociated products of digestion in an anti-normal state of being will then be consumed, whereupon that which we commonly call 'life' unfolds like lightning.

With this final realisation I was overwhelmed by an irrepressible urge to write everything down and as far as possible to build the requisite experimental devices. This I bequeath to all those who make this mysterious journey through life after I've gone. In the light of this new knowledge they can spare themselves the trouble of believing what our erudite idiots or sophisticated boot-lickers teach in schools and churches.

Excommunication – and Little Wandering Springs
From *TAU* Magazine, No.147 (July 1936).

Today I received a warning from friendly quarters that many exposés of mine were provoking a great deal of irritation in certain circles which would not permit themselves to be exposed to ridicule in such a manner and would in no way take it lying down. I replied that it was far from my intention to expose anyone to ridicule: I considered the present state of affairs much too serious. That the articles I have written to date have seriously harmed me financially, I freely admit and I do not think the worse of anybody who may feel affected, when they defend themselves as best they can. When I say "*as best they can*", in no way do I sanction the exploitation of positions of authority for unfair means of defence as has frequently occurred, but that any defence should be so formulated that scientific circles should have no difficulty in refuting these attacks factually or at least in mitigating them. Should this be impossible then even deceit will serve no purpose. Countless individuals have already become aware of and are beginning to reflect on the many inconsistencies in contemporary scientific interpretations. If these often totally perverse practices continue to be applied, it will very soon thoroughly wipe the smiles off all our faces.

As Zimmermann[42] so aptly informed rowdy students at the time...."*things have already gone so far today that the will is missing even to listen to another point of view, so then those of us, who for purely personal reasons, try to suppress essential solutions vital to our existence, are able to cause even more harm.*" In my opinion my adversaries get so annoyed because in many instances we are concerned here with things which are so simple and natural. One has to scratch one's head as to why those, who devote their whole lives to the clarification of these natural phenomena and who are supported by the public purse to the end of their days, were themselves not aware of them. One-sided activity apparently produces mental sterility and thus the very expression, 'an opinion lacking any scientific basis' provides grounds for gravest concern.

It is really quite unnecessary to go right back to Noah, who became the first marine officer from sheer horror of water, in order to establish that there may be other sources of meaningful, economic and social facts, apart from qualified authorities. In many instances it can actually be an impediment to be possessed of too much so-called knowledge. The Maid of Orleans succeeded in winning battles regardless of the entrenched theories of strategy and tactics. In the same way it should be possible, despite the resistance of feudal knowledge, to solve the enormous natural, scientific and technical problem, which indubitably is waiting in the wings. All the elements were always there although much has been covered up, by people who believed they could step outside the natural order of things.

The remarkable aspect of the course of my life is that it was the disputes themselves, indeed the positive attempts at my elimination, which always took matters a gigantic step forward. Therefore I will mention two conversations that occurred at the time of my defensive battle. The first conversation was with the former Minister for Agriculture, Andreas Thaler, and the second with the now deceased Prime Minister, Dr Ignaz Seipel.

When I was the duly appointed, albeit unenviable, water-wizard of the Federal Austrian Forestry Department, no week passed without an uproar in one province or another only later to subside in some minister's office. Thus one evening I was summoned to the ministry in order to give an account about some incident or other. At the time, as far as I recall, it was about a general attack by all the forest administrators who demanded my most expeditious removal from government service.

[42] Professor Werner Zimmermann, publisher and editor of the ecologically oriented magazine *TAU*, who published a number of articles by Viktor Schauberger and was also Viktor's stalwart friend in adversity.- Ed.

The Prime Minister, Dr Seipel, greeted me with the words, *"Tell me, what is actually the matter with you? Every week you are brought to my attention. What weighty matters did you discuss with my Minister of Agriculture today which made him send you over to me immediately?"*

"Prime Minister, perhaps you may still recall that without any action on my part, I was urged by Mr Loew, the head of department, and Mr Buchinger, the then Minister of Agriculture, to leave private employment and enter government service. The reason for the appointment was a log-flume I built a few years ago which, against all expectations, achieved extraordinary savings in running costs. These savings were made possible only because I had come to understand the true nature of water. In my view the current practices of modern water-resources management should be prohibited by law as quickly as possible. Not only will all the waterways be ruined, but agriculture as well, the groundwater table will sink and the time will soon come when the whole production of high-quality food in Central Europe will be jeopardised.

"The forest will also die if all clear-felling operations are not forbidden immediately, for through the denudation of the forest floor, ecologically-essential differences in temperature will be lost, which enable vital calciferous substances[43] to enter the interior of the tree. What contemporary forestry regards as light-induced growth is in reality one of the symptoms of cancer-promoting disease, because the enlargement of capillaries disrupts the processes of diffusion necessary for the organic growth of the tree. Owing to the resulting increase in oxygen pressure, symptoms of inflammation must appear which in my view are precursors to the emergence of cancerous decay and the qualitative demise of the high forest.

"The present system of artificial fertilisation is extremely dangerous because the soil will be ruined and the infiltrating slag-residues will block up soil capillaries, resulting in the cessation of interactions between geosphere and cosmos. Terrestrial earth-rays are bound by these energetically-leached residues and only draw inflammation-enhancing calcific[44] substances from the soil, leaving inflammation-impeding calciferous substances behind to form hard-pans.

Prime Minister! Earth, water and air are organisms with mutually contrasting potentials and are able to transform and build each other up reciprocally only as long as humanity does not disturb these naturally-ordained reciprocities that have functioned for millions of years, reciprocities which actually order life itself."

"Are you Catholic?" Seipel then asked. *"Yes, Excellency, although I must mention that according to the information passed down to me by my father, our family is still burdened by a 12th century papal excommunication."* *"What do you mean?"* asked Seipel. *"My ancestor, so the story goes, was a man of long-standing nobility who revolted against the temporal seat of the Archbishop of Passau. He was captured and shortly thereafter his height was shortened by a head in accordance with*

[43] 'Calciferous': Calcium-forming or -producing substances, especially calcium carbonate. – Ed.
[44] 'Calcific': Forming lime or chalk. – Ed.

the decrees of a papal bull." Seipel stared at me for a long time and dismissed me with the request that I should speak further with Minister Thaler. Thus I soon found myself once more in the presence of this illustrious peasant Minister, who amongst other things, related the following experience to me.

"In the vicinity of my farmhouse in Tyrol runs a spring, which I always observed with much anxiety during hot weather because this little spring was the be-all and end-all for myself and my farm. In this way, year after year, I was able to establish that this little spring discharged higher and higher up as everything around it began to dry up and turn brown. The hotter it was, the colder the water became, and eventually during an extremely dry summer it emerged about twenty paces higher up. The water became better and there was more of it. Water really harbours so many secrets", stated the intelligent minister, *"and I fear that in many things you are perhaps quite right. However, powerful forces are arrayed against you because you hold completely different views and are not easily overcome."*

Many efforts by way of anonymous documents failed to compensate me for the unpleasantnesses I suffered. Shortly thereafter my employment as water-wizard came to an end. To some God gives in sleep, while others must learn with greater difficulty through the fickle fortunes of life The course of my later life took me to many nations and thus I came to the barren mountains of Montenegro, a region of particular interest to a water researcher. One day my thirsty mount led me to a remarkable spring at an altitude of about 1,200m. After a protracted time communicating in sign language, the accompanying Muslim explained that this spring constantly wandered. In hot weather it emerges high up on the mountain, and in the cold season it flows in the valley where, apart from other related phenomena, its discharge diminishes.

At the point of entry into the light, this spring was exactly +4°C (+39.2°F). About 15m downstream it was already about +8°C (+46.4°F). The most remarkable aspect of all was that at this lower point I could measure an almost threefold quantity of water, although the channel fell almost vertically and any possible inflow did not exist. Any kind of braking of the water was also impossible because the water flowed over completely smooth rock. For the first time I was confronted with a practical example of *water-growth* and also with a wandering spring similar to the one that Minister Thaler had described to me.

Apart from the growth of the water itself and the striking increase in temperature, no other form of growth was evident at the higher position, whereas at the place where the spring flowed during the cooler times of year a burgeoning profusion of all manner of aquatic plants was to be found. If at the upper position water was drunk directly through the lips, then I experienced a conspicuously strong dizziness, almost akin to an inebriated state, whereas the water at the lower point, which I also repeatedly visited at other times of year, produced no ill-effects. The fact that the spring stemmed from one and the same source was proven by the drying up in the lower position at which, despite the lower external temperature, the spring was warmer than it was when it emerged higher up on very hot days.

I was able to establish the reason for the peculiar behaviour of the water only years later. Air and water are the products of intermediate spaces, *space-lattices*,[45] in which a large variety of diffusive functions take place, wherein the effects of heat or cold are significantly altered. The greater the resistance offered to the Sun's rays, the finer the space-lattice mesh becomes and the colder the water and vice versa. If the pores in the stones lying on the bottom of the spring are closed as a result of cold influences exerted externally, then in spite of very low external temperatures the water will become warmer. Springs which exhibit temperatures of about +4°C (39.2°F) during the hottest periods warm up to about +7°C to +8°C (+44.6°F – +46.4°F) at cold external temperatures of about -32°C (-25.6°F). Alpine huntsmen call these springs 'warm waters' and they set their fox-traps near them, because these springs never freeze; on the contrary, they become increasingly warmer the colder it is outside.

For a long time I was unable to understand why the rock in the vicinity of the spring was so homogeneous, despite direct irradiation by the Sun. Eventually this mystery too was explained. In a certain sense, such rock not only grows with direct sunlight but it also contracts its pores mechanically, in order to protect itself from the effect of the Sun's rays[46]. The smaller the pores, the cooler the emerging water. This is because only the most ethereal, high-frequency rays[47] are able to penetrate the rock. It is this quantitatively-reduced but qualitatively-increased solar pressure which cools the warmest water by many degrees within a few seconds. However, if the pores of a stone lying in a good spring contract, due to the physical influence of cold then, as mentioned earlier, the water will become warmer since in this case only the most noble [refined] geospheric rays are able to emerge and combine with concentrated *oxygenes*[48].

All growth is founded on this wonderful Law of Reciprocity,[49] and thus we find that in contrast to shade-demanding species of timber, light-demanding timbers have a completely different bark structure. The latter rely on thick bark as a means of protection whereas the former produce a dense crown-closure in order to flourish. It is therefore quite self-

[45] 'Space-lattice': An energetic atomic matrix or structural configuration. – Ed

[46] 'Ethereal high-frequency rays': The higher the frequency of the radiation, the less its expansive, heating effect. For example, ultra-violet light is a high-frequency radiation whereas infra-red light is a lower frequency which at its lower levels becomes thermal, heat-producing radiation. – Ed.

[47] 'Geospheric rays': Nurturing female radiant emissions stemming from the Earth. – Ed.

[48] 'Oxygenes': The various grades and categories of oxygen. – Ed.

[49] 'Law of Reciprocity': The law governing the inversely proportional interaction of opposites. – Ed.

evident that any high forest will inevitably die if young, shade-demanding saplings (such as spruce) are exposed to direct sunlight through clear-felling.

The skin-colour of different human races, and the various types of animal fur are also to be attributed to this phenomenon, which reflects the indirect processes of growth[50]. This alteration of the pore morphology gives rise to changes in general metabolic activity, wherein lie the origins of the scourge of the twentieth century – cancer. In order to make this terrible plague disappear again we must take heed of the intermediate processes of growth which are directly connected with these phenomena, wherein energies are diffused indirectly. For this reason the simplest remedy for cancer is *water* infused with the most highly refined ethericities of a geospheric nature.[51]

Above all, this water possesses a remarkable property. Shortly after drinking, it warms up, and about two hours later it generates a singular coolness in the body which noticeably refreshes the whole organism. Apart from raising sexual potency, it also has quite a remarkable effect mentally, and therefore after a longish period of consumption even thoughts that are able to penetrate these secrets are produced. In this way the veils concealing Nature's many secrets automatically begin to fall away.

The Sun's rays, falling on surrounding protective rock, will homogenise its fabric through the deposition of transitional substances. The same sort of deposition occurs in the body when sun-drenched water is drunk constantly. Conversely, when we drink water that has been correctly constituted geospherically, or if we take in foodstuffs that have evolved under this natural inner protection, then dissolution of the transitional substances takes place. A phenomenon of particular note is that under the influence of light, good springwater produces only bacteria essential to life. High-quality fish are attracted to it in the spawning season in order to provide their young with the noblest of fare.

These processes of *ur-genesis* are indirectly connected with the rock formation surrounding the spring because the peak effects of the mutually opposed terrestrial and solar rays can only give birth to the most highly evolved entities. This explains why high-grade bacteria are present only in water in which inferior substances are incarcerated, as it were, in the surrounding material. This is constantly patrolled by policemen (free, dissolved oxygenes) in order to prevent the emergence of the above lower-grade life-forms.

Thus in all water or blood-supply vessels in which the contents are conducted in an organically-correct fashion, we find that dangerous opposites

---

[50] In the case of human skin colour, this relates to the different ways in which sunlight, for example, is diffused through the variously coloured pigments as a result of the differences in colour. – Ed.

[51] 'Geospheric': Belonging to the Geosphere and originating from within the Earth, be it of material or immaterial nature. – Ed

are separated from each other in one and the same profile. This enables interactions between individual peak-effects to take place incessantly. These are responsible for the growth of water or the quantitative increase of blood, provided the necessary sterilising mechanisms are present and the external tissues are in a healthy condition.

Nature does marvellous things whenever she decides to unveil her mysterious ways to humanity, the crown of creation. She directs and protects those who honour and faithfully serve her, and for this reason I have cast to the four winds even the most well-intentioned warnings of concerned friends. Life is a remarkable illusion. It is ever in the service of one form of metamorphosis or another, and therefore even a papal excommunication can be transformed into a blessing if omniscient Nature so desires.

Return to Culture
Part of an article written in 1932, the remainder unfortunately being lost.

Our accustomed way of thinking, in many ways and perhaps without exception, is opposed to the true workings of Nature. According to our way of thinking it is inexplicable why cold water is unable to accelerate when flowing down an unobstructed sloping course into a valley. Science does indeed provide explanations for this which, however, do not apply. Likewise, how is the warm, lighter water of the Gulf Stream able to displace cold, heavier, surrounding water, and in a broad, deep current travel across thousands of kilometres, even in a reverse (uphill) gradient?[52]

Equally incomprehensible is why cold, heavy groundwater high up in the mountains does not sink downwards, and under certain circumstances it even rises. Or conversely, why is it cold in the stratosphere and why is warm, light air unable to rise upwards? Here the resistances against its rising appear to be significantly less than inside the Earth, where water, despite its own weight and at times with difficulty, can still wend its upward way.

No-one who has ever seen a riverbed in the Sahara (a *Qued*), which in times of drought is a gigantic flat, yellow trough full of gravel, could imagine that in the event of a sudden rainstorm cold water cannot penetrate

[52] 'Gradient': In terms of Viktor Schauberger's concepts, a gradient refers principally to temperature. Temperature is seen as a condition of energy, and with a decrease in temperature (positive temperature gradient), water gains in density, dynamics, potential and energy content as it approaches a maximum at +4° Celsius, its so-called *anomaly point*. In the present instance the gradient can also be interpreted in a topographical sense, in that the seabed under the Gulf Stream slopes upwards from the Caribbean Deep to the Dogger Bank, so in a fashion the Gulf Stream is actually flowing uphill. – Ed.

into the hot earth. This can happen with such severity that shepherds with their whole flocks, and troops of soldiers, man and horse alike, can be drowned.

More puzzling still is the phenomenon that when approaching the Sun it becomes colder, and conversely, that the temperature rises with increasing distance from the Sun. If the Sun is at its zenith (perihelion), when we experience its greatest heat, the explanation is again lacking as to why, when the Sun is furthest away (aphelion), the most intense radiant energy is encountered.

As it flows down a steep incline, even a cold, clear forest stream presents us with an apparently almost insoluble enigma. The more precipitately it flows down a steep slope, the less it damages its bed and banks. If the same stream becomes exposed due to deforestation, and if its water is warmed by the direct light of the Sun, it becomes lighter and less dense. It then immediately begins to destroy its banks and becomes a raging torrent.

Mosses demonstrate a law to us, which is initially incomprehensible to our accustomed way of thinking. In the cold, fast-flowing water the tips of mosses growing on the stream-bed and on the slippery stones play a remarkable game amidst this rushing turmoil. Should the temperature of the water approach +4°C, when it becomes heavier and more solid, its flow faster and more torrential, then the tips of the moss point *upstream*. Curiously enough, if the water is warmer and therefore lighter, and its relative forward motion is slowed down, then these same moss-tips point *downstream*.

However, not only do these moss-tips seem to defy known laws, but so does other life natural to this environment. The fish, the mountain trout which lives in such rushing forest streams, reveals to us wonder upon wonder. The colder, heavier, swifter and clearer the water becomes, the more motionless the trout's stance and the faster it is able to flee upstream when danger threatens. If however water is exposed (through removal of the protective tree-cover) and warmed up by the now unobstructed incident rays of the Sun, it becomes lighter, more turbulent and its forward motion is impeded. Under such conditions it is impossible for the trout to maintain its station in the flowing water without physical effort. The trout is unable to maintain its former almost motionless, tranquil stance – its characteristic and natural right, but through activity becomes visible and swiftly falls victim to its enemies. The trout now begins an ever harder struggle for its very existence. As the water warms its food is ejected from its formerly prescribed path where once, without any effort on the part of the trout, it was swept into its jaws.

In exactly the same way that our thoughtless interference has made the trout's existence more precarious, our incorrect way of thinking has also made the struggle for our own survival more difficult. Our thinking is inconsistent with what we actually see. The eye is a perfect, natural organ. The seen image is a reaction phenomenon. Using an artificial optical apparatus, the same effect can only be obtained in a roundabout way, by means of a negative. The eye, on the other hand, immediately presents us with the diapositive, which is the true image.

Our sight constitutes an unconscious, automatic transformation process. Our thinking however is really a purely individual, conscious and, under certain circumstances therefore, a learnable process. If our *thinking* is to attain the same perfection as our *seeing*, then we must change our way of thinking and learn to see reality not as an *action*, but as a *reaction*. Perfect thought lies in the apprehension of the correct reaction, for before the eye can show us the positive, it must first transform the negative. In effect it must break up what it records. What we see therefore, is the turning inside out of what we receive. What our mind grasps in this way must be re-formed and re-thought if we wish to attain that for which we strive.[53]

There are many contradictions which appear to lie between our thinking, our action and what exists in reality. One, however, on closer examination threatens the whole edifice of scientific development. As we shall see later, this phenomenon is proof that even a stone should be viewed as an *organic structure*, a living system, and that what we call 'the Sun' is also the same as the Northern Lights. Both are *phenomena arising from cold oxidation* precipitated by *low temperatures*, and have nothing in common with the concept of fire and of direct heat.

Two completely identical pebbles or two identical, organically grown and flawlessly structured high-grade timbers, when rubbed together under water, can produce a clearly perceptible fiery glow or ray of light. It is indeed curious as to how *two completely identical objects* can produce fire when rubbed together, and the question arises: *what is actually burning*? Why isn't this fire extinguished by the surrounding water? For on the contrary, the radiance actually intensifies when the water approaches its anomalistic temperature of +4°C.

We cannot be concerned here with heat due to friction in the commonly accepted sense, since frictional heat cannot be created or maintained under

---

[53] "It has been proven psychologically that human beings can only appreciate or apprise, i.e. comprehend and understand, something new, if they can succeed in raising into their higher consciousness the subconscious immured in their brain cells. If this cannot be achieved, then all preaching is useless. And even the eye has first to learn how to see everything new; it too must therefore be aroused from its latency before it can grasp the seen. Above all, there must be a readiness to consider even supposed wonders as the forerunners of forthcoming realities, for only thus can the rational mind be provided with a basis for calculation and analysis." From V.S. letter to Hermann Jaeger, 31st October 1957. – *Implosion* Magazine, No.103, p.20.

water. We are therefore confronted with a phenomenon which is impossible to explain in terms of current scientific theory. *De facto*, no branch of science, with its present makeshift concepts, is in a position to explain the cause of this effect. Not only this phenomenon but many other surprises arise out of a closer examination when viewed from our accustomed standpoint. These inevitably lead us closer to the realisation that our way of thinking, and thereby our way of acting, stand in opposition to the workings of Nature and are therefore against Nature.

In fact, on closer inspection it can be shown that our increasingly widespread impoverishment is principally to be traced back to gross flaws in logic; to errors which in part had their beginnings thousands of years ago, and to which the wealth of whole nations have already fallen victim.

Thus we increasingly come face to face with the conscious realisation that in many cases we have built on *false theorems and principles*. Unfortunately, these false principles can neither be removed from the biological catalogue, nor from humanity's memory. Passed on as they are from generation to generation, they always remain the basis for our economic legislation. Since it is founded on false premises, it must inevitably lead to the economic decline of those nations deemed highly cultured by today's concepts.

In the following, as a typical example from school will demonstrate, one of the greatest errors in thinking which has, astonishingly enough, been able to persist for decades, leading a whole world astray. When the farmer clearly understands what water actually is and what role water and its habitat, the forest, play in the whole economy of life then the present activities of our forest and water industries would be *forbidden by law.*

As our chemists tell us, water is H_2O. The fact that water can be decomposed into its component parts, H_2 and O, with the aid of an electric current has served as proof of this for decades. Practical experiment demonstrates, however, that distilled or chemically-pure water, pure H_2O, cannot be decomposed by an electric current. That is unless acid has first been added to it, in order, it is said, to be able to conduct the current. After the electrolysis[54] has been completed, what remains behind is once again water, H_2O. If the experiment with electrolysis is to be repeated successfully, then acid must again be added to the water. This is a sure sign that acid is no longer present in the residual water, in other words, that it has been used up in the process.

Should one of the products of decomposition, so-called hydrogen, now be burnt (hot oxidation), then the obligatory residue, water, is again present, and in addition, carbon dioxide (CO_2), which is associated with all combustion processes. The remaining water, or the hydrogen that was 'burnt' to

[54] See Viktor Schauberger's article "Electrolysis" in *The Fertile Earth*, vol. 3 of the Ecotechnology series. – Ed.

water, was therefore only the *carrier of carbones* (fig. 4), which were restructured (gasified) by the electric current. Such carbones, for example, are present in household well-water, and in the case of distilled water are introduced into the water by the addition of acid.

Fig. 4: Hydrogen, the carrier of oxygen and carbone.

In practical terms, we are not concerned here with the combustion of hydrogen gas, but with the combustion of carbones present in the hydrogen (the carrier H), from which ultimately all acids are formed. These carbones combust with the aid of higher temperatures, or at appropriate low temperatures undergo a process of cold oxidation. The essential, crucial and previously unknown aspect of the matter, however, is that it does not involve any chemical bond between O and H. Instead, under certain movements of temperature (temperature gradients) *oxidation* occurs between O and the carbones concealed in the carrier H, which leads to the carbones' remodelling, enhancement, or transformation.

Quite apart from other natural conformities with law associated with it, this process of qualitatively up-grading (ennobling) the carbone totally departs from accepted theory. It must take place, however, for the very reason that the *immediately-following oxidation* always happens under a *lower* temperature than the preceding one. This occurs if the process is allowed to proceed in the proper way and is not derailed by the interference of people and other disturbances. If this oxidation takes place under the right temperature gradient, then apart from the reconstitution (qualitative ennoblement) of the carbones, there is a simultaneous charging of energetic potential, which increases the level of energy in the material itself. This leads to increased potential and tension in and between the new substances. The increased build-up in energetic charge and potential always occurs in proportion to the work done through the oxidation. The stronger and hence more naturalesque the oxidation, the higher is the quality of the product of this process, namely the substance.

We are therefore confronted by the fact that up to now our view of energy and the way it is maintained is incorrect, and that as a result *action* and *reaction* do not represent *a balanced system*. The reaction is a multiple of the action, provided the preconditions for the reaction or oxidation take place according to law, i.e. if O and C are correctly distributed in the carrier H, and the oxidation can take place under the right temperature gradient. With this realisation alone all our economic concepts to date are truly turned upside down, since our present economic thinking is focused on 'taking' and pure mechanics (direct action). It is thus unsuitable for any practical economic activity. All the real phenomena of Nature, for example, the increased growth of vegetation, which is ultimately all we strive to emulate, are only *reactions* (indirect effects).

The present economic situation is the logical result of this serious error in thinking. The apparently unavoidable economic decline, however, is only the thoroughly befitting consequence of a total failure to perceive the laws prevailing in Nature. It is a result of completely erroneous interpretations which, quite logically, lead to further mistaken activities. These mistaken activities (our work) must legitimately lead to increasingly widespread unemployment, because our present methods of working have a purely mechanical basis. They are already destroying not only all of Nature's formative processes, but also the growth of vegetation itself, which is being damaged even as it grows.

The logical outcome of the disturbance of Nature's most fundamental creative processes can only be impoverishment due to our meddling. This decline, the only tangible effect of our utterly misdirected work, however, is a necessary precursor for reconstruction. Without grinding poverty and a solemn warning sign we shall never come to our senses and learn to understand that Nature has no sense of humour, when we clutch at her innards with greedy, clumsy hands. Nature, disturbed in her development by our mindless endeavours, will only then reliably and legitimately permit us a rehabilitation and reconstruction, the day we break with the present system and its concepts. Instead, with our God-given eyes, we need to base our future actions on more thoroughly considered notions which are seen to have their foundation in reality.

The following practical examples will demonstrate that we shall have to accustom ourselves to fundamentally different attitudes. We will have to base our present way of looking at things, our scientific and economic laws and as a priority our whole educational system, on completely different principles. This is imperative if we wish to halt the total economic collapse steadily gaining ground, and find our way back to culture.

The sum of the various forms of mechanical and metaphysical[55] energy is not constant. If we view matter as concentrated energy[56] in accordance with Einstein, and make the logical assumption that matter will be of a higher order the more energy appears to be concentrated in it, then the purpose of

---

[55] 'Metaphysical': here the German word is *psychisch*, which can be interpreted as *psychic, metaphysical* or *of the nature of the mind*. 'Metaphysical' has been chosen in preference to 'psychic' as it seems more in keeping with what is discussed later on. – Ed.

[56] In this context, energy is actually not to be defined in terms of the accepted concept, but is only here retained here for the sake of general understanding. – VS.

this accumulation can only be the development and the *qualitative increase of matter*. All increase in the creation of matter is initiated by *thermal motion*. As is well known to us through scientific experiments, every process of transformation is associated with differences in pressure as a function of thermal motion in the carrier of the process.

It is impossible for like to beget like. The above transformation can therefore only be brought about through the interaction of opposites. We are thus compelled to assume that this thermal motion was itself the opposite of another motion, for otherwise neither motion nor transformation could occur. The picture of motion is somewhat relative, for it is immaterial whether the displacement of two points in relation to each other is due to a change in the position of either one or the other (like the motion of the landscape passing by the window of a travelling train).

This thermal motion, however, requires a carrier, which must be something indivisible, i.e. an element (though not in a chemical sense). The causal agent of thermal motion can thus only be the product of other opposites which are present in a common carrier – the element hydrogen, to which the general role of *carrier* of the process of transformation has already been attributed.

Heat is not identical with energy, but only a manifestation of its function. Once again it is impossible for energy to arise from itself, from its like, for energy is the external result of the inner potential or charge between mutually repellent elements of carbone and oxygen particles in the common carrier hydrogen. Upon closer examination both these elements cannot in the long term remain hostile to each other, because they seek to unite with each other in matter, and thereby to shape the matter into a more highly developed form. This union once more generates a mutual impulse to separate, and through such separation arises the will to reunite (tension). This interplay of reciprocal separation and reunion now leads to the increased accumulation of energy. Through this more highly organised matter develops, which represents a resultant energy incorporating a higher number of interacting oscillations.

If in this way the will for reunion is proven through the fact of separation, then it is impossible for these opposites to effect a state of absolute inactivity. However, they do provide our sense of perception with an apparent picture of peace, in which their individual phases are no longer visible to us due to the high rate of vibration or oscillation. (The number of cycles per second required to maintain a steady incandescent electric light amounts to roughly 50 hertz). Here we have the concept of higher development. If the purpose of the development of higher structures is not to be negated by the highest exponent of evolution, namely humanity itself, then it is impossible for Mayer's 'Law of the Conservation of Energy' to continue to exist in its present generally accepted form.

In accordance with the way in which physical processes are presently studied, only the process itself and its various stages are observed. It is assumed that the carrier of the process remains unchanging, whereas in reality it also changes. This also relates to Mayer's principle of energy. Where only the conditions for equilibrium in mechanical forms of energy are considered, it is doubtlessly correct. Silence, however, is maintained about metaphysical forms of energy. If Mayer's Law of Energy proves the conservation of mechanical forms of energy, then when considering metaphysical energy the conclusion is reached that there must be a further development, a continuously evolving growth that we can actually see in Nature. Therefore, if we include metaphysical forms of energy in our consideration of equilibrium, then we recognise that in this case action and reaction do not remain equal. Rather, that under certain circumstances, the latter may be a multiple of the former, even though we have as yet no yardstick for the possible comparison of both processes in terms of units of measurement.

The collapse of our present world-view permits the thought to surface that our whole way of working, even in its conceptual development, is incorrect. To influence the less highly organised natural processes in agriculture, forestry and water resources, the bases for increased growth, in the right way (and not to 'correct' them wilfully as we have done thus far), it is necessary to acquaint ourselves with the most fundamental mechanism for initiating activity. In other words, our thought processes and how they come into being. As we know from scientific research, every thought process is associated with differences in heat and blood pressure in the various parts of the body, as well as with other processes of which we are currently unaware. Through differences in temperature and the differences in blood pressure thus created, the heart activates the blood circulation and the delivery of those formative substances required for the growth of the body. Naturally, it must also supply the brain with nutrients, which are transformed there into higher grade products in order to produce thoughts.

This transformation, however, can only occur when, after the accumulation of sufficient energy, the brain itself becomes an epicentre of energy. It is then able to transform energy from one form into another independently, in order ultimately to be able to emit it as a ray, radiance or information. This epicentre is then itself a source of energy in which oxidations can occur. A thought process, whose manner of coming into being has so far remained unknown to us, is the creation of a thinking apparatus that extracts its transformative substances from the blood circulated by the action of the heart. The process of thinking is thus a form of oxidation which, through the appropriate accumulation of a minus charge (oxygen), permits the brain to become a secondary focus of energy that is in a position to transform energy into thoughts and to radiate them. This activity takes place in the convolutions of the brain, which have the structure of capillaries and thus act as resistances.

The steeper and more tightly wound the convolutions of the brain and the more numerous the capillaries, the greater the transformation of the nutritive substances and thus the higher the quality of thought. The energies emitted from the brain in the form of radiation actually have no carrier; they are carrier-less. The proof that radiation is carrier-less is demonstrated by the way that radiation in a vacuum is not perceptible, whereas it is perceptible in a body due to the presence of the carrier H. Here the radiation experiences resistance and is transformed either into light or heat, in order finally to reproduce itself in wave form.

These transformation processes create the structure of the brain. Similar transformation processes are also decisive in the creation of the remainder of the animate and inanimate world – for example, that of the tree and the stone, which for the following reasons primarily interest us here.[57] Agriculture and all the other politico-economic factors related to it, are dependent upon the correct growth of forest. In trees too, the higher the charging and radiation of matter, the smaller the capillaries in which the oxidation takes place. These important capillaries have been destroyed by humanity's interference through the light-induced stimulation of growth in shade-demanding timbers. This will inaugurate the decline of the forest industry, a decline that will be hallmarked by a degeneration in quality.

---

[57] Since it relates to forestry, the final section of "Return to Culture", entitled "Forestry Agriculture and Wood Production", has been included in *The Fertile Earth*, vol. 3 of the Ecotechnology series. – Ed.

The Genesis of Water

The Genesis of Water
From the Schauberger Archives – Viktor Schauberger, Linz, January 1952

In the French monastery of Arles-sur-Tech there stands in a cool crypt a great sarcophagus, made of the finest marble. For the last 700 years or so, there has been a fresh and powerful healing water in this sarcophagus, with which thousands of people with incurable diseases have been able to regain their health. It is a healing spring of a type similar to the world-famous healing source of Mary of Lourdes, which after much discussion and investigation was pronounced a wonder by the Church. The monks of this Pyrenean monastery offered a prize of 1000 gold francs for the solution of this enigma, and were the mystery to remain unsolved, then without doubt, the water springing from this sarcophagus would receive a halo.

Every high-quality spring that arises from the gigantic sarcophagus of the Earth is the greatest wonder; a wonder, however, which is known to dry up if the mouth of the spring is exposed to the direct light of the Sun. Under these circumstances up-welling water becomes specifically lighter, and because of this it ought to spring much higher above the ground were it dependent on pressure alone, which is the conventional scientific explanation of this phenomenon.

We, however, will attempt to unveil perhaps the most profound of Nature's secrets, the *ur-genesis* of the Blood of the Earth. In so doing scientific knowledge, which lies an octave too low, will be raised, thereby attaining a level on a par with that of the *ur*-ancient *Religio*, with the higher order of knowledge of earlier more highly-civilised cultures. These ancient peoples understood how to make use of the *inner* creative and uplifting forces required to inaugurate the *ur*-production of material substance, of all growth. Undoubtedly they also made use of them as the almost cost-free drive for powering specially constructed machines. Ancient people were of such high spiritual standing that they also succeeded in overcoming physical gravity with the aid of the *original* motion of water.

This *original* motion also accounts for the autonomous formation of sap and blood and its *original* circulation without excessive hydraulic pressure.

The purpose of *original* motion is the concentration of *dynagen* from which, conditioned by reversionary cosmic influences, a growth unfolds, which is developed, multiplied and qualitatively improved to one stage higher. This marvel, the true perpetual motion of Nature cannot be explained purely speculatively so here we must take examples from Nature. One of the most informative is the dewdrop, which is supposed to stand on the tips of grass. It also explains the apparent wonder in the Pyrenean cloister of Arles-sur-Tech and above and beyond this, the marvel of the creation of all that we see around us.

The dewdrop is the physically *first-born*, which the indirect light of the Sun imbues with new life, and whose direct rays endow it with *ur-form* (prototype), with body and soul. Thus the foundation stone for development of new life is laid, without which there would be no reproduction and propagation, no multiplying and ennobling of older forms of evolution. There would be no development on this terrestrial manure-heap Earth, in whose interior high-grade fermentation processes take place, unless the owners of forest, farm, water and energy were as cautious as the doctor, who carefully ensures that no free oxygen enters the veins through his hypodermic needle. The reason for this precaution is best explained by the doctor, who has become wiser through unfortunate experiences (as far as he is capable of doing so), for even the most naturally aware of doctors is unaware of how sap, juices and blood are actually created.

Early in the morning when the dawn begins to break, the atmosphere becomes noticeably chillier; the temperature approaches the anomaly point[58] of +4°C (+39°F). When the external temperature reaches this condition of neutral stability and the state of *indifference*, there appears on the tips of the grass a very delicate protoplasm-like tensile form, which similar to a soap bubble or a small air-balloon, stands motionless on each tip. When the rising Sun shines upon this primal, elemental and untrammelled form, the protoplasm fills itself up to two-thirds with water. With, and because of, increasing heat from the Sun, the intensity of the light decreases[59] and the grass tips supporting the increasing weight of the dewdrop succumb to the law of gravity and bend down towards the ground. Finally the little womb-like sack bursts and the juvenile, *ur*-begotten water trickles into the Earth, accelerated by the blade of grass as it straightens up.

[58] 'Anomaly point': The behaviour of water differs from other liquids. While all liquids become consistently and steadily denser with cooling, water, alone reaches its densest state at a temperature of +4°C (+39.2°F). This is the so-called 'anomaly point'. This Viktor Schauberger termed its state of 'indifference', which in his view is decisive in terms of its potency and its influence on quality. It is also the temperature at which water has its greatest energy content, least spacial volume and is virtually incompressible. Above and below this temperature water expands, hence its anomalous nature vis-à-vis other liquids. – Ed.

[59] Blue light is high-frequency, high intensity cold light, therefore greater warmth is manifested as light frequencies reduce towards infra-red. – Ed.

If one wishes to infuse a sick and tired body with the constructive, creative and uplifting energy concentrated in this *ur*-form, then one must walk on these protoplasms as soon as possible before the impending birth of the *ur*-water. For once the water is born, the intensifying life-force in this minute sarcophagus is already committed to this physical first-birth; in other words, the new water-body has been united with its soul. This materialisation only ensues when, in the anomaly state, the fertilisation of a negatively-charged concentration of *dynagen* by a bipolar counterpart can take place through a process of diffusion. Even then, it can only take place when the soil is metalliferous and trace elements with a specific charge or valency are able to play their part. These trace elements serve as a spiritual-energetic connecting bond or a catalyst for the resultant emulsion (the intimate and intense union between bipolar counterparts).

Were the catalyst – the copper pipe – in the sarcophagus at Arles-sur-Tech to be removed, then this would probably signal the end of all genesis of juvenile blood of the Earth within those marble walls. This pipe apparently displays no trace of oxidation whatsoever, despite hundreds of years of storage, due to the absence of free oxygen in this hermetically-sealed container. The pipe would oxidise and the water would vanish, however, if the crypt was heated and if atmospheric oxygen, by this means becoming aggressive, was permitted to penetrate into the interior.

Just how crucial is the anomaly state, not only for the crypt itself, but also for the interior of the coffin, is demonstrated by any wrought iron nail. For example, when embedded in a wooden sole-plate immersed within the anomaly zone of the groundwater, it exhibits no signs of decay (rust) whatsoever. Here too there is no release nor an increase in the aggressive behaviour of oxygen, which ur-generates either creative, uplifting energy or decomposive energy.

In such a manner, even a grain of corn can maintain its fertility over thousands of years if it has lain in correctly and naturally acclimatised burial chambers or royal sepulchres in which it is protected from the influences of free oxygen. This protection is afforded by the artificially created anomaly zone present in these tombs or burial mounds, which have been constructed in a naturalesque way which inhibit putrefaction. This proves that those prominent in religion and society, who arranged to be buried in cool vaults within coffins made of certain alloys, were well aware of the difference between putrefaction and decomposition. The decomposition of the blood and bodily fluids takes place immediately when free atmospheric oxygen succeeds in entering a vein. It also happens when latent oxygen in a sap-duct becomes aggressive through over-exposure to light or overheating (in the case where shade-demanding timbers are planted out in the open) and binds the fructigens, which become passive under these temperature influences. The origin of fructigens will be more precisely explained later.

We are therefore concerned with fundamentally different fermentation processes, which, controlled by temperature, generate development-enhancing or development-impeding forms of energy. The former lead to the precipitation of amniotic fluid and subsequently to im-plosion, which functions in a cell-building way. In the latter, the decomposition and decay of the physically first-born results. The concerted effect of such decomposition leads to reflected radiation of emanatory essences (horizontal ground rays), which, like an electric current, also decompose the surrounding groundwater. This leads to oxyhydrogen gas-like *ex*-plosions, once the bipolar basic elements have been separated. More details about this will be given later.

Incidentally, the *levitational* current, arising from the preceding *im*-plosion event, possesses contractive, biomagnetic forces. In exactly the same way as the Earth's magnetism, these act along the longitudinal axis, and in naturally-flowing water, flow against the direction of the current towards the origin or source of the water. For example, any trout standing in this high-grade concentration of dynagen is drawn along in their wake. By means of its adjustable gill system, this fish is in a position to manipulate this levitative force so that it either stands motionless amidst the rushing springwater, accelerates upstream, or floats upwards in the middle of freely-falling watermasses. This happens if the falling watermasses, whose flow is determined by suction-curves developed prior to the actual waterfall, can rotate around themselves, within themselves and about their own axes in cycloid-spiral curves. In the process an ideal (subtle-spiritual) axis can come into being, out of which, as occurs in our blood stream, the hitherto unresearched emanatory essences[60] flow, which imbue the body with life. To the extent required for its freedom of movement, a trout makes use of these essences to overcome its body-weight effortlessly. This is conditional on its blood maintaining the requisite anomaly state, which is the condition of indifference (neutral potential) and thus freedom from fever. In other words, the trout is healthy and thus capable of reproduction and deployment.

All formation of water, sap, juice and blood takes place by way of diffusion. This is the reason why a doctor is careful when giving an injection, so that under no circumstances may free atmospheric oxygen enter a vein. In such an event, conditioned by the blood temperature alone, decomposition of the blood can occur.

With this the supposed wonder of the genesis of *ur*-water in the sarcophagus mentioned earlier now becomes understandable. The formation of high-quality (cool) sources of water in plants and trees takes place in the same fashion.

So-called 'savages' puncture plants to slake their thirst with this exceptionally wholesome water, thus becoming so intuitive or receptive towards

---

[60] In function and character these are akin to 'ethericities'. See footnote 20, p.27. – Ed.

external influences that they become spiritually aware. Understanding how to make use of this supposed wonder, they stand far above civilised people, who are led astray by relying solely on their speculative talents. As a result, these poor-in-spirit, greedy little misers become more and more spiritually sluggish, utterly ordinary speculators and ponderers, who are incapable of understanding highly-intuitive people. Indeed they view such advanced thinkers as madmen. In actual fact the spiritual dimwits are the real 'lunatics', having created this fool's paradise of ours in which we can just about manage to vegetate. In our degenerative way of working, we not only devalue our food but also saw off the very branch of Life on which we sit. Of itself, this can overcome not only physical, but also spiritual gravity. After this digression, these supposed wonders should be objectively and matter-of-factly described. Whereupon every kind of hypocrisy will become superfluous, indeed ridiculous.

A decade of observing the trout standing motionless in up-welling spring water has given me an insight into the deepest of Nature's secrets, so the theme of this apparent mastery of gravity should be more closely discussed.

The first and most important condition for this phenomenon is that downward-flowing masses of springwater are able to in-wind themselves mechanically. At the same time, through a decrease in temperature towards +4°C (39°F), these watermasses are able to become specifically and physically denser. For this a particular configuration of curves is necessary. These we find especially well-portrayed in glacier *moulins*,[61] which even in the hardest rock are shaped by correctly adducting water[62]. In a similar fashion such water can also form blood and sap vessels. If the surrounding rocks are alloyed with the appropriate minerals, becoming metalliferous, then the first two preconditions have been satisfied. The falling water can then convolute in cycloid-spiral space-curves about itself, within itself and around its own ideal axis in opposite directions in the *original* manner of planetary motion. Thus a biomagnetic (contractive) dynagen concentration is created in the middle of the flow. In exactly the same way the metaphysical (physically imperceptible) Tree of Life, whose branches grow out into physical reality, comes into being along the biomagnetic axis of the Earth.

Only by means of the interplay of mechanical and physical in-winding, which causes freely-falling water to approach the anomaly point, are biochemical products of reaction made possible. Viewed purely functionally, here we are referring to 'levitational' force, the opposite to that of physical 'gravitational' force'. Under naturesque preconditions, these two forces are so balanced that gravitational force predominates. But only to the extent necessary for accelerated motion (speed of fall) of masses due to specific

[61] Vertical shafts in a glacier maintained by a constant descending stream of water and debris. – Ed.
[62] The process of drawing or swirling inward into the central axis of its own wake. – Ed.

densification (approaching the anomaly point). However, in the opposite direction, the intensity of levitational force thus triggered off increases to the extent required to maintain (at an almost constant level) the steadiness in pace of free-falling watermasses or watermasses flowing down a variety of gradients.

In other words, the specifically heavier the naturalesquely falling or flowing water, the greater the levitational counter-current. Through the interaction of counter-flowing emanatory essences, this counter-current projects itself upstream and manifests itself as *new water*. This explains why naturalesquely falling or flowing watermasses can grow or increase. At the same time the watermasses are also qualitatively improved due to a heightened generation of levitation energy. In all processes of growth, heat-substances (latent oxygen = solidified solar energy) are consumed, so the additional *new water* masses thus created are also caused to approach the anomaly point. This produces additional formative and uplifting energies, which also undergo a quantitative and qualitative intensification. The intensification of these levitational energies in the longitudinal stream-axis becomes so powerful that the heaviest trout is sucked into their wake and is then drawn upwards. Due to this effect these fish, inwardly aroused, can climb at full speed on their nuptial path towards the source-spring every spawning season. This explains the phenomenon of the stationary trout, which overcomes not only its own bodyweight, but also simultaneously, the gravity of the water flowing against it.

But we must return to the discussion of water itself in order to unveil the secret of what is actually to be understood by the expression 'binding the Sun's energies', through which the *implosive* force evolves. Implosive forces will make all present atomic energies (the lower grade *explosive* forces) completely uninteresting. The over-exploitation of coal, timber and oil will then cease automatically. In terms of the concept of 'energy', we need to understand the finely balanced self-equilibrating action of a bipolar dynagenic structure, which, depending on the imparted impulse, precipitates either *fire* or *water*.

Since long-term economic growth, let alone a higher level of culture, cannot be achieved using the first of the two dynagen emulsions, *e*xplosion is excluded at the outset. In this case oxygen is consumed (burnt). Under the influence of heat oxygen is freed and becomes aggressive, whereas the *fructigen* of opposite polarity, a latent fatty substance, becomes passive (inactive) under the influence of higher temperatures. This subject will be discussed in more precise detail later. Here we are primarily concerned with the description of a higher-grade fermentation process, which becomes active when falling or flowing water becomes specifically denser during its descent, thereby accelerating and simultaneously approaching its anomaly state.

Water flowing or falling in in-winding (involuting) cycloid-spiral space-curves never freezes. It bores almost silently into the pool below and in order to wind itself upwards again, it extends its radius of action by rotating about itself, within itself and about its own axis. Viewed as a whole, an *ur*-works comes into being which, like a clock spring, is driven by an increase in weight (water-increase plus specific densation). The eternal state of rest-lessness, the *panta rhei*, is thus attained. It is the *original* transformation-serving reproductive and propagative motion, the same type of motion of sap and blood, which is divorced from any pressure-gradient.

In this system of circulation which takes place in cycloid-spiral space-curves, it is the *tempo-curves*, whose shape induces frictionless motion, which are responsible for the constant acceleration of descending water. Again, by means of contradirectional brake-curves, the steadiness in the flow of such water is achieved, regardless of the geological gradient; its individual boundary velocities can neither be exceeded nor decreased. The essential aspect of this *original* flowing motion is that, where the continuously alternating temperature gradients intersect, whose nature is determined by the type of motion, the development-enhancing energies are freed and the counterforce responsible for steadiness of flow becomes active.

In this way the standing or flowing blood of the Earth – water – becomes a sort of sarcophagus. Out of this a bipolar pattern of emanations streams forth in all directions (precisely determinable), penetrating everything in the surroundings. Through the influence of high-grade dynagens, it animates and energises the environs. At the same time however, retroactive, repulsive forces become free and active, eliminating everything capable of impeding or otherwise harming this process of further development.

In naturally flowing, simultaneously increasing and qualitatively improving water, regardless of external temperature influences or changing conditions of gradient, *im*plosive as well as *ex*plosive processes take place. These can be so ordered that the formative and levitative products of reaction are principally employed in bringing the surroundings to life. This inner metabolic cycle is therefore the life-motor – a four-stroke motor in which actively impanding and expanding, repelling and thrusting motions rhythmically alternate. A pulse-beat arises here, which is not the *ur*-cause, but the outer effect of an inner reciprocating process.

If one throws a piece of wood into a properly functioning *ur*-works (such the pool at the base of a waterfall), then one can clearly perceive cycloid-spiral space-curve water-mixing processes taking place. Here we are concerned with a peculiar swaying movement, an 'elfin dance', which proceeds in the same way as a dance in which bipolar sexes (man and woman) move around in an irregular circle, while simultaneously gyrating in waltz-like fashion about their common axis. This peculiar round dance is also perfored by spawning trout coming up from below as they prepare to float upwards within the middle of the waterfall.

In the same way that a human dancing couple take the floor and seek the moving beat of the music, the trout also orientates and adjusts itself to the pulse-beat of the upward-circulating water in the pool. Suddenly it begins to sway in snake-like fashion at the edge of the pool and in-winds itself towards up-winding watermasses. It finally reaches the place where the waterfall plunges into the pool. It then dives in the direction of the axis of the falling water and begins its run-up for the intended ascent. In powerful looping movements it seeks and finds the source of the levitation force, which begins at the interface between the falling and upwelling water. It is now drawn into the vortex of the reactive up-current, which it assists through strong movement of the gills, like a pike before lunging like lightning at its prey.

With good illumination, the path of the levitational currents is visible now and then. It appears as a tube, apparently empty of water, which we can perceive when falling water circulates with a gurgling sound above a drain, forming a downwardly-directed whirlpool which, with increasing suction, drags everything with it into the depths. Were we now to imagine this whirlpool or water-cyclone operating in the *opposite* direction, then we have a picture of the reactive upsuctional effect. The phenomenon of the trout floating upward in the axis of fall, which has so far been a puzzle, has thus been solved.

This floating upwards by the trout is connected with very particular dynamic processes in falling water. In just the same way the activation of a water-spout or a cyclone is also associated with very pronounced differences in potential which can only occur in the tropics. If falling water cannot in-wind itself mechanically and physically, then the biochemical products of reactive substances cannot come into being either. The forces they produce possess a superordinate (metaphysical) upsuctional power, which in the above case draws the trout upward in its wake.

The original cause of this cyclonic force is a high-grade fermentation process, which can occur only under quite specific types of water flow. These are types of flow which trigger implosion, the binding of the stock of latent oxygen by naturalesquely-fermented fatty substances (carbones). With regard to the term 'oxygen', it has to be understood here as *solidified solar energy*. It becomes so inactive at +4°C, the anomaly state of water, that in a mechanically atomised (dopey, drowsy) state it can be bound (consumed) by its counterpart, which in the anomaly state has become free and monopolar. This reactive product of dynagen has an *implosive* formative and levitative function. In this way it is possible to increase the *ur*-production, which will lead us out of the fool's paradise into which we came by way of unnatural methods of increasing productivity.

The Increase in *Ur*-Production

Today there is constant talk of increasing production and productivity. Productivity of this kind is achieved at the cost of *ur*-produced biomass. This declines both in quality and quantity if the preconditions for *ur*-production and growth are disturbed through destruction of the nutrient-processing zones of the soil and through unnatural and incorrect methods of water-retention and supply. Instead of *procreative* (high-grade) fermentation processes, warm interactions are triggered off. In an over-heated or over-illuminated cellar, for example, these give rise to fundamentally different products of fermentation compared to those that arise in a fresh (cool) cellar. In the latter a high-grade product is developed through the interaction of emanatory essences (reduction processes), as happens in the crypt of the monastery mentioned earlier. The anomaly zone, so crucial here, also occurs in watercourses whose motion is technically contrived naturalesquely, regardless of constantly changing atmospheric influences. This means that the anomaly state, so decisive for all growth and for the 'fever-free', healthy reproductive and propagative status of water, can be almost constantly maintained in winter and summer.

To make this clearer, this will now be explained in more detail. When the upwelling watermasses in the waterfall pool overflow the crest of the containing sandbank, they again flow in a geological gradient and will continue to inwind if two space-curve flow-guides are placed immediately downstream. These impart a naturalesque impulse to the current, the effect of which is maintained for at least 10km (6 miles).

Despite these precautions, since the draining water is usually exposed to strong solar irradiation, it exhibits a noticeable reduction in the carrying capacity and tractive force. These are instilled into it in the vicinity of the spring and cannot be maintained unless 'cooling stones' are placed in it or are already present. When draining water collides with well-formed, metalliferous riverbed stones, the laterally-deflected water-masses are imparted an additional rotation about their own axes. This again induces a cooling towards the anomaly point (+4°C/+39°F), which is so important for growth and development. Measured at right angles to the axis of flow, the reduction of temperature varies from 0.1°C – 0.4°C (0.18°F – 0.72°F). If it is taken into account that an expenditure of energy in the order of 42,700kgm is required to heat up one cubic metre of water by 0.1°C (0.18°F), then per stone, this represents an increase in bio-energetic potential of about 85,000 kilogramme-metres (kgm). Therefore enormous constructive energies are lost if incorrectly regulated watercourses are warmed up to normal bathing temperature (about 20°C/68°F)) in high summer, or freeze in winter. Due to the *original* flow-motion, which maintains the anomaly condition, both of these conditions are impossible in naturalesquely regulated waterways, despite the influence of high external temperatures.

In those places where stones resist motion, the additional forces mentioned earlier are freed and become active. Moving three-dimensionally in the *original* way, the water also becomes denser. The suctional force centred in the axis is also intensified, producing an increase in tractive force and carrying capacity by which sediment is transported and gradually rubbed together without heat. Bodies heavier than water are drawn into the locations where suction vortices arise. Trout also stand here, for their food swims effortlessly into their jaws. Only on very sultry and thundery days does the latent oxygen become warm. The condition of flow also deteriorates, resulting in a slight reduction in the tractive force and transportive capacity of the water thus expanded. In this instance trout are also forced to hunt for their irregularly distributed food, now mostly stranded on the banks as flotsam and jetsam. Because of this the fish, as anglers say, 'begin to bite', for they are becoming hungry. Prof. Dr. Forchheimer's surprise was proof of the extent to which this is still unknown. Through his careful measurements of temperature, this world-famous hydrologist became convinced of the significance of the effect that resistances in the shape of well-formed and alloyed stones have on flowing water.

A temperature change of only 0.2°C (0.36°F) in a fertilised chicken's egg is enough to inhibit the quickening of the living embryo within, resulting in the generation of rotten gas. Similarly, a few tenths of a degree difference in the temperature of flowing water determines its capacity for reproduction and improvement in quality. It also determines the possibility for the increase and ennoblement of the physically first-born – water. No wonder that, as a result of a complete disregard for *original* flow-motion, today's hydraulically regulated water decays and dies. Instead of non-pathogenic bacteria, pathogenic bacteria which are dangerous to health come to life.

Naturally-flowing water is therefore a carrier of health, whereas when beginning to stagnate, it becomes the epicentre of an epidemic and a lethal danger to human and beast. It is also to be noted that temperature influences arising from moving water are more powerful than those of the surrounding water and air temperatures. Whether an increase or decrease in temperature occurs in the course of flow, is therefore dependent solely upon the type of water motion. This determines which process of fermentation, warm or cold, will prevail, which in turn determines whether a formative or destructive dynagen-concentration is brought into being. The basic principles and fundamental precepts to which every water supplier should seriously pay heed, follow from this.

Every form of water movement that triggers an increase in temperature (fever) in the accelerated watermass is just as developmentally dangerous as a motion triggered by temperatures. It is immaterial whether the motion-inducing temperature influence lies above or below the anomaly point. Therefore only those types of water movement which accelerate water-masses artificially come into question. Whether this is in a river, turbine, pump or canal is unimportant. What is needed is water movement which approaches the condition of health (+4°C/+39°F), for only then is economic growth possible. The natural, legitimate and progressive consequence of today's customary methods for mass-movement of water (conventional hydraulics), for tilling earth (by plough), pumping water (by pressure turbine and pressure pump) or air-movement (propeller and pressure screw), are pressure and heat-increasing processes of retrogressive development (cancer). This will debase all growth, *ur*-production, the more intensive and widespread the present methods of working become.

Temperature and the Movement of Water

Every movement of water will trigger off reactions if it contains the bipolar (counter-potentiated) basic building blocks of life in a latent transitional state. An equivalence (transformation of motion into equal values of heat or *vice versa*) is therefore just as impossible as a constant maintenance of dynagen. Which type of dynagen concentration arises is dependent upon fermentation processes previously triggered by the gentlest of motive impulses. According to the type of fermentation, the dynagen emits life-giving or life-taking matrices of emanatory essences in the form of radiation.

When, for example, a life-*giving* dynagen concentration produced by *original* mass-motion (where the temperature approaches +4°C, the anomaly state of health) emits radiation, high-grade fermentation processes are generated in the surrounding body of groundwater affected by such radiation. Through these fermentation processes, higher-grade products of reactive substances are also created, simultaneously leading to the formation of juvenile (new) water, in which the freed energies are accumulated and bound. In other words the dynagen concentrate, this higher-grade emanatory substance transmitter, comes into being, which functions at a level one octave higher. This then gives rise to new chain-reactions in the surroundings, which have a life-giving function.

If, by comparison, water is made to move hydraulically (pressure-increasing and heat-inducing), then precisely the reversed course of development occurs. For in this case water-decomposing radiation is emitted, which functions in exactly the same way as electricity when conducted through water. That is to say, a decay of basic substances takes place, whose synthesis results in detonating-gas explosion and a reactive development of heat. This explains how a river or a current that descends at a rate of 800m³ (28,250 cu ft) of water per second can heat up to bathing temperature (about 20°C or 68°F) in high summer.

In order to heat up one cubic metre of water by only 0.1°C, an input of energy in the order of about 42,700 kgm is required. If this amount of water is heated to bathing temperature, it can only happen through the agency of the reactive development of heat mentioned above. In this way the higher-grade, superior-quality energies are lost, which serve to build up the vegetation. In the above case of 800m³ of water per second (the average flow of the Danube or the Rhine), this signifies an energy loss of about 60,000,000 hp/sec or 45,000,000 kW/sec. Were water moved in the opposite way to that prescribed by today's authorities, all this lost energy could be harnessed towards building up the vegetation.

This simple calculation shows just how far-reaching are the mistakes of such people as Robert Mayer. Mayer, ignorant of the reactive consequences of every type of water motion, postulated the concept of equivalence and the Law of the Conservation of Energy arising from it. The same is also true of Isaac Newton's Law of Gravity,[63] which failed to take into account the *development of levitational power*. Without this there could be no build-up of weight (no growth) and therefore no constantly increasing body-weight. The combined effect of these two laws would mean that there would be absolutely no motion, but rather an eternal developmental standstill and therefore no life on this terrestrial dung-heap, Earth.

The Earth's own *original* motion is triggered off by a higher-grade interaction between basic elements, which can only happen through the potential differences arising where light and heat are excluded. For this reason alone all over-illumination and over-heating triggers over-acidification, and through this the processes of decay and retrogressive development (cancer), all of which are the result of excess oxygen. This oxygen becomes all the more aggressive as the Earth's root zone (anomaly zone) is heated up through reversed reactive effects. An epicentre of epidemics is created, resulting in the impregnation of the *ur*-production with the seeds of decay, which it then transfers to human and beast by way of polluted foodstuffs.

Only a small mechanical, physical or psychic impulse or agitation is all that is required to trigger off the appalling scourge of this technical, hydraulic or dynamic age amongst flora and fauna. Present day researchers of cancer have also overlooked this communicative process. Consequently and in accordance with Nature's laws, more animals and human beings must prematurely decay alive, the greater the successes achieved by tech-

[63] Viktor Schauberger commented wryly on the subject of Newton's discovery of the Law of Gravity: *"I think it would have been much better, had Newton contemplated how the apple got up there in the first place!"* (*Implosion* Magazine, No.35, p.16). – Ed.

nology, hydraulics, dynamics and chemistry. These disciplines have jointly overlooked the degenerative and reactive consequences of all forms of accelerated motion which deviate from the *original* form of motion in earth, air and water.

But to return to the actual matter in hand; any water caused to move unnaturally or stimulated by temperature expands both in summer and winter. Only in very rare cases does it attain its anomaly and healthy state, where it is can reproduce and build itself up (increase and improve itself qualitatively) and is once more able to release evolutive emanatory ethericities into the environment; so to breathe life into it and cause it to move in the *original* manner.

Water forced into unnatural motion loses its specific density and thus its carrying capacity and tractive force. It deposits its sediment in which are secreted latent stocks of raw materials and basic elements. Under natural conditions of flow these elements rub against one another without heat and by the indirect route of further fermentation and germination processes, they furnish the increasing watermasses with the nutrients contained in this water-sarcophagus. This cannot happen unless the anomaly state ($+4°C$ to $+9°C/+39°F$ to $+48°F$) of the water is maintained by the mechanical and physical in-winding motion of the draining water. Above and beyond this, and as a consequence of *warm* fermentation processes, the build-up of a decomposing form of dynagen throughout the entire ur-production already begins. This manifests itself in erosion of the riverbank, pot-holes, one-sided deposition of sediment, silting up of the watercourse, flooding and droughts, in the degeneration and death of both water and more highly organised creatures; and lastly in the emergence of pathogenic parasites.

If the foundation stone of evolution degenerates and dies, then throughout the whole course of development a process of regressive development (cancer) is set in motion. In forestry its beginnings are exhibited in 'light-induced growth', the coarsening of the molecular structure of the *ur*-production throughout the whole spectrum of growth. Therefore what the forest owner, that artificially-fertilising, clear-felling and excessive exploiter of forests, regards as a scientific achievement: supposedly accelerated growth, is none other than a rank proliferation of cancer. The roots of this cancer lie in already-degenerated cell-space through which it receives continuously-increasing and intensifying decomposive forces. Hence the incurability of cancer; small *ur*-causes, but great effects. For all life arises out of the *original* motion of water, which is characterised by a decrease in temperature in the direction of the anomaly point. In no single case has this so far been achieved with contemporary methods of conducting water, whether in open spaces, artificial channels, pumps or turbines. Indeed, precisely the opposite has actually been achieved.

Flowing Biomagnetism

In terms of this so far unknown concept, one has to imagine the exact opposite of so-called *'electricism'*. Up to now it has not been recognised that electricity is the reactivated (and physically imperceptible) *pressural force* which arises out of over-illuminated and overheated cell structures through the agency of warm fermentation processes. This pressure-imparting electrical force is the origin of the rank cancerous growth and coarsening of tissues mentioned above. The counter-concept of *biomagnetism*, representing cell-forming, fabric-condensing (contractive), formative and levitational forces, must also be more closely defined. At the same time the principal building block of Nature (the vital, finely dispersed residues of earthly fatty matter) will be examined. Its natural reaction-product demonstrates the essence of all growth and, coincidentally, the wrong track that modern science has so far trodden and which we have to thank for causing the constantly increasing mass-misery throughout the whole civilised world.

It was a forester, ambushed and killed by poachers, who showed a living forester the way in which flowing magnetism could be generated mechanically. In the *Frankfurter Illustrierter* on the 6th of May 1951 (No 18) a short article appeared, stating that after four years the body of a forester, who had disappeared without trace, had been discovered in a potato patch near the edge of a forest. An observant police officer had come across a place about 1 metre wide, where the potato plants were noticeably darker and their growth more luxuriant. Here they dug and found the body of the murdered forester in which the bullet, the catalyst, was still embedded (viz. the copper pipe in the sarcophagus at the monastery of Arles-sur-Tech). Many years ago in the region of Nuremberg, the grave of a murder victim was also discovered in whose pockets there were a few coins, and whose corpse was interred under a large and very leafy tree.

Over thirty years ago I came across a flattish mound in a virgin, alpine forest, upon whose rich greenness flowers of superb colours blossomed that strikingly did not belong to the surroundings. I once spent the night in this remote spot, in order to be close at daybreak to a gamecock's courtship display ground. Towards midnight, just where this curious mound was situated, a bluish-white flame licked upwards. My first thought was that I had thrown an unextinguished match onto this moor-like spot and sprang up in order to put out the incipient forest fire.

In the meantime the flame had grown a metre high and took on an egg-shaped form, similar to those that now and again issue from rock fissures and like shining dewdrops, stand on the point of a rock. Many years ago a chief forester from Vienna, Walter Hackel, photographed just such a strange light over a metre high. A copy of the photograph was unfortunately lost at the end of the war, or rather was stolen by looters of my apartment.

However, at that time I knew nothing of these things and so I backed away in horror, as I stood in this pitch-black darkness before an ever more powerfully flaring and heatless flame, which threw a pale glimmer into the surroundings. At first, like a man possessed, with my heavy mountain staff, I hit at the place from where tongue after tongue of this mysterious egg-light sprang up. When I noticed that this shaft of light only sprang from the rock at one point, I ceased to flail away at the supposed forest fire and loosened the surrounding soil. This, however, changed nothing. Then I held my hand in this egg-light and instead of feeling the anticipated heat, experienced an icy coldness and saw the bones standing out on my hand. An icy chill ran down my back. I returned to the tree where my gun lay, released the safety catch and sat down in my former bivouac and waited to see what else would happen. After about two hours the sky at last began to grey. A few hundred metres away the gamecock began his courtship, the actual reason for my early visit. I didn't move from my position, watching how this uncanny glow slowly extinguished, and suddenly the whole spectre was over.

When at last daylight came, I returned again to the source of the flame and on every tip of the lush green leaves I saw oversize dewdrops, again in egg-shaped form, standing motionless like glittering candle flames. As the first rays of the Sun pierced the tangle of leaves, the grass-tips bent under the weight of the *ur*-water, which visibly grew as the Sun's heat increased. One by one, the now finished dewdrops fell down.

Now I began to dig into the hillock with the tip of my mountain staff and underneath a peculiarly smelling layer of humus, I felt a resistance, which after further digging, turned out to be the almost undecomposed corpse of a chamois buck, which had a clearly distinguishable bullet entry hole above the left foreleg. There was, however, no exit hole. According to the time of year, it could only have been shot by poachers, since the hunting season was long past. It was only later that it became clear to me that the buck must already have lain underneath this mound for a longish period, because it was covered with a thick layer of humus upon which vegetation had doubtless sprouted from time to time. On even closer inspection, I found a sort of mass-grave before me.

The old hunters used to insist that chamois (as also happens with elephants) seek out special places to die where slow decomposition rather than putrefaction takes place. Sick wild animals are attracted to such places which remain equally warm or cold in winter and in summer, seeking either a cure or a painless death. Expressed scientifically, constant anomaly states prevail, which permit decay-free decomposition. This is why, as a particularly sly old forester explained, the high clergymen had themselves buried in a constantly cool church crypt, or why the more common priests at least had a little roof built over their graves along the cemetery wall at the eastern side in order to protect them from rainwater. I realised later that, because of its free oxygen content, which activates decomposive forces, rainwater actually promotes decay or rusting.

Eerie lights also dance over such graves from time to time and on nights of the full moon, upwelling skeins of mist are now and again visible. Only about 40 years ago in the Aigner Cemetery (Mühlviertel), this triggered sheer panic amongst the populace who saw the dead priests doing an elfin round dance.

These strange wafts of mist that occasionally ascend from close stands of mixed conifers at about 11 o'clock in the morning are known to every forester and are a reliable sign that there will be a heavy thunderstorm the same afternoon. Above the Nuremberg gravemounds on the eastern side of a thickly wooded slope, similar phenomena are well known. At excavated grave-sites the bodies exhibit so-called 'soul-holes' above the chest. These are always missing, however, where the bodies have been burnt before burial, as indicated by smoke blackened stones surrounding the skeleton.

A geologist and well known gravemound investigator by the name of Kerl explained that the latter cases related to those people whose physical and spiritual energies were considered undesirable for the afterworld and therefore received neither a soul-hole, nor a cool burial place. This geologist also contributed to the discovery of the above murder victim, because one night on his way home he saw a light eerily dancing over a cool shaded spot. In the *Wiener Erzähler* a longish article by the physicist Lormand appeared under the heading "Mysterious lights demystified!", in which he referred to the physicist Mestelle and his forerunner Dumas, whose experiments in part explained this phenomenon.

"At certain places on the moor strange lights (will-o'-the-wisps) suddenly began to dance, seriously alarming the people. Excavation demonstrated that in these places a wild animal had been slaughtered. Mestelle then had about 30 cow skulls systematically buried in which decay had already set in, for in his opinion the brain mass, which is especially rich in phosphorus, was the source of these uncanny lights. About 11 months later weird lights began to dance over these burial spots, convincing Mestelle that he had finally unveiled this mysterious spook. He let it be known that right up until the Last Judgement, when their verdicts would be pronounced, the souls of the dead would rest in peace. 'Honest people would not be permitted to incandesce as they decay', hence we are here concerned merely with the self-ignition of decomposive gases."

That which burns, heats, and what heats has its origin in annihilating fire, which arises through the agency of a warm fermentation process. Genuine will-o'-the-wisps radiate no heat, but a conspicuous coolness. They only occur where the anomaly zone prevails beneath the surface of the ground.

Proceeding with this line of thought, and after many observations of naturally flowing rivers, I recognised the origin of the well-known gleaming of the so-called 'Nibelungen' or 'Rhinegold'. Only occurring in naturally flowing streams, this glow is emitted by mature limestones as they rub together in the coldest possible water. This ultimately led to the discovery of a mechanical way to generate upwardly-flowing biomagnetism, or levitational force, along the longitudinal axis of a river, which effortlessly overcomes all weight and manifests itself outwardly as temperature-less light. Such light is also known to be generated by certain types of fish deep in the ocean, merely by moving their bodies in cycloid-spiral space-curves. This takes place in the negatively-charged abyssal anomaly zone, which for thousands of years has imparted the quickening impulse to the source of all life and the whole manifested world.

And what does science have to say about future developments?

In fifty years' time the atom bomb will be uninteresting for the technologist. The sources of coal and oil, which by then will also have been exhausted, will likewise be totally uninteresting, for power and light will then be generated by solar energy. It will not be possible to do without fermentation processes altogether, however, even if they are used merely to improve the taste of alcoholic drinks! Without fermentation processes there is no bipolarity, no potential difference, no motion and no life. Of course there would also be no water, no Blood of the Earth, the *ur*-source of all we can see around us or otherwise perceive.

Modern science seems to have misconstrued cause and effect, otherwise scientists could neither speak of just *one* type of energy, nor could they exclusively consider it to be the only existing functional source of the bipolar concept of *motion*. Motion, however, is an interplay between opposing bipolar elements, upon whose influence the type of motion that comes into being depends. Every motion consists of *components* of kinetic energies. In scientific terminology these are called *compressive* and *tensile components*. It is the components of the pre-active and/or predominant motive force operating in a given motion which determine whether life-*giving* or life-*taking* dynagenic effects result. These effects are always initiated by either warm or invigorating (revitalising) fermentation processes, since a dynagen concentrate is only a limited conglomerate of raw material, whose effect can only be triggered by a motive impulse.

Gunpowder is another such amalgam of raw materials. Its explosive effect stems from its latent dynagen concentrates and is caused by some motive impulse. The same also applies in the case of very fine flour or coal dust, which due to the presence of atmospheric oxygen, explodes when it is stepped on with a metal-studdied shoe. For this reason, only felt slippers may be worn in gunpowder and explosive factories. Despite this, several powder factories have already blown up!

This provokes the question: what therefore is solar energy? It is doubtless only the *heat* generated when super-positively charged oxygen consumes (binds) its counterpart, the reflected radiation of the terrestrial dynagen concentrate.

The counter-question arises: what happens when this emulsion, which arises from the intermixing of counter-polar emanatory essences, takes place in reversed order? What happens when naturalesquely-fermented carbone (mistakenly called hydrogen), which in the anomaly state becomes free and monopolar, binds (naturesquely consumes) the mechanically most finely dispersed, sluggish oxygen? The answer to this is missing in modern scientific encyclopaedias, because it is still unknown that such a thing as a (predominantly) *negatively*-charged dynagen concentrate also exists. Here yawns a great gaping hole in our knowledge which first has to be closed through the study of naturalesque processes before solar energy can be exploited on a large scale technically or more correctly, eco-technically.

If bipolar emanations of dynagen are therefore *intermixed* under conditions hermetically sealed from external influences, then a predominantly *im*plosive effect results, as occurs for example in sap and blood vessels. However, the *ex*plosive effect is also present as it has a role to play in the necessary excretion of faecal matter. Exactly the opposite occurs in the case of a purely technical, hydraulic or chemical generation of motive power. Here, without exception, the reactive oxidising effect is always uppermost, which is dangerous for all forms of evolution. Conforming to natural law, a retrogressive development is thus inaugurated due to the lack of counteracting reductive effects, the result of which is cancer.

If the wonder of the Pyrenean monastery mentioned earlier cannot be explained, then it is simply because the origins of the build-up of higher-grade juices and of the formation of blood, have so far not been recognised. Today it is still believed that the heart is a hydraulic pressure pump driven by muscular contractions. Every body of whatever kind, whether mineral, metal, plant, animal or human being, is none other than a sort of sarcophagus, a diffuse resistance, required for the intermixing of emanatory essences.

If one lays a stone or a board on the ground, moisture develops underneath it. When this resistance is correctly diminished, then between earth and stone or board a negatively charged anomaly condition is created, which even denies atmospherically-influenced rainwater access to the Holy of Holies (the anomaly zone). If the weather becomes sultry and excessive heat is exerted on the now well developed anomaly zone under the hot stone, then earthworms begin to migrate by the thousand. As do these earthworms, the stationary trout in rivers also perish, because the preconditions for the existence of these noble fish can only be found in a well-developed anomaly zone. A general rise in sea temperature is also responsible for the migration to cooler zones of some sea-fish; a fact already recognised by science.

Were the crypt in the monastery at Arles-sur-Tech to be heated, or the interior of the sarcophagus altered through the influence of atmospheric light or heat, then the marvel of the growth of water would cease. Indeed all that would perhaps be required to end it would be to allow free atmospheric oxygen constant and unimpeded access to the interior of the sarcophagus, or to remove the copper pipe, which would very soon show signs of oxidation.

Today's technology, hydraulics and dynamics utilise degenerative forms of light and heat in mass-produced, unnaturally shaped and alloyed mechanical monsters, which necessitate the combustion of stocks of raw materials (crucial for natural processes of evolutionary development). Moreover, the resistances associated with these processes also have to be overcome; resistances that increase by the square of the increase in velocity. As a result one must ask in all seriousness whether one is not here concerned with exceptional fools.

If in contrast, one considers the almost silent operation of *im*plosion machines, which transform existing resistances into additional power, then one understands how a multiplicity of reactivated matter can be transformed into a higher quality life-form by means of additional reactive forces. These reactive forces are available in almost unlimited quantity, due to higher-grade fermentation processes produced by the naturally-fermented fatty residues of deceased life-forms in the anomaly zone. This also explains how, from the minutest seed, a gigantic tree can arise, which itself creates a thousand million seeds. A properly planted potato can produce up to 20kg (44lbs) of high-quality potatoes within half a year if one merely remembers to incorporate the type of negatively-charged catalyst required for every process of transformation. In other words, the catalyst must be present that produces an emulsion made up of combinations of emanatory essences of contrasting nature, through which water is precipitated. Thanks to the discovery of the *im*plosion process, such water can be *ur*-created mechanically in any amount and of the best quality almost cost-free. Thus it will be possible to make the boundless desert regions fruitful once more.

The Dethronement of Science

From *Implosion* Magazine, No.50, (written by V.S. in May 1955)

The teachings of the *Tabula Smaragdina* were incised into the hardest precious stone – the emerald. Handed down to us from antiquity, they preach that a healthy and happy existence is dependent on the naturalesque intermixture of the stuff of Heaven and Earth. The progeny of this marriage between exalted atmospheric and geospheric ethericities is *water*, the blood of the Earth. This *ur*-source of life comes into being when the ethericities of the Earth bind those of Heaven. For this the maternal forces and energies must be more powerful than the incident fertilising substances, for if the process takes place in reverse order, then *fire* is created.

The art of increasing and qualitatively improving water has been so thoroughly lost over the millennia that today no one knows what water actually is. How it comes into being, how it deteriorates or how it dies and subsides is still a mystery. If the degeneration and disappearance of water cannot be successfully halted, the Earth will turn into a treeless desert. The whole of civilisation is threatened! The survival or disappearance of humanity is thus at stake. We are concerned here with neither politics, nor racism, nor an invention in the usual sense of the word, but with the question: *what is more important for evolution and development, water or fire?* This is what it is all about, and nothing else! Everything else must retreat far into the background. Without water there is no habitat and no life!

Most people know that machines are indirectly driven by fire. Entirely unknown, however, is that energy can be achieved far more simply and cheaply through the artificial production of *water*. Furthermore, growth can be promoted, increased and ennobled in the process. Therefore not only will the problem of an independent existence be solved, but in the future there will be no shortage of raw materials provided the mixture of heavenly and earthly essences can be naturalesquely fostered. This involves the industrial duplication of 'photosynthesis', which is conditional on the possibility of energising the hydrocarbones. With this discovery (which the University of California has so far sought in vain) all armed conflict will be unnecessary. In future no sane individual will allow himself to be drawn into a war merely to satisfy power-mongers.

All robbery will likewise cease once everything is available in superabundance, for with the artificial production of water, vast tracts of desert can be restored to their former fertility and to provode living space for the ever-increasing mass of humanity. There will also no longer be the need to exploit this ancient art for purposes of war. Whether the disturbance of the water-balance happened out of ignorance or to maintain the economic *status quo*, through the iniquitous world-wide traffic in foodstuffs and raw materials, cannot be addressed here.

The fact is that, as today's technology has progressed, the point has been reached where the proliferation of cancer is not only universal, but apparently unstoppable. For this reason the only solution is to put an end to the self-deception by reintroducing practices that promote the natural treatment and care of water. All this has nothing to do with science for it involves the diametric reversal of the methods applied by today's fire-spitting technology, a technology that should be considered the true causative agent of cancer.

Perhaps it is necessary for the misery and privation of humanity to become so intense, before it is finally recognised that science, as the guiding force of humanity, has made a disastrous blunder by implementing reversed energetic interactions between basic elements. In consequence, any possibility of resuscitating Nature's principal formative substance, reduced protein, was brought to a halt. It was thus inevitable that the processes of decay would become so rife in both air and water. This exemplifies the *disastrous errors in motion* and excitation made by a science founded on *unnatural principles*.

The Nobel prize winner, Otto Warburg, proposed in a scientific journal that the formation of cancer cells was caused by a deficiency of oxygen. A healthy cell is supplied with oxygen by the blood and uses it to process nutrients, thereby acquiring life-energy. However, Warburg has now asserted that the cell receives too little oxygen. As a result the metabolism is reversed and the lactic acid becomes thoroughly fermented, the cell grows and divides more rapidly and in the process turns into a cancer cell. This is also confirmed by Professor Domagk. Both researchers, however, made a crucial mistake. They failed to see that it makes a considerable difference whether *modified* or *unmodified* oxygen enters the blood. Modified oxygen is formed when it is filtered by the lungs. Only then can it be taken up by fructigenic ethericities via the bloodstream, which on their part have been purified by the filters of the intestines.

If this process of purification is in any way disturbed, then an oxygen deficiency results, leading to suffocation and putrefaction of the unfertilised ethericities of protein. This is the answer to the riddle. This is where a yawning gap in scientific knowledge lies and without its closure cancer cannot be cured. Nature precludes the use of one and the same form of energy for different purposes.

Hence the report continues: *"Those afflicted by cancer and whose lives are threatened by this terrible disease will read of Warburg's findings with resignation. Moreover, they will demand to know why this oxygen deficiency has not been prevented. According to Warburg, in this way the incidence of cancer could be reduced to a fraction of its present state if the cells of the body could be protected from chronic damage. How this was supposed to happen, however, he was unable to say."*

Dr Liebmann, director of the Biological Research Institute in Munich, held that this involved the disturbance of the biological equilibrium. He overlooked the fact, however, that a state of equilibrium would necessarily imply a *developmental standstill*. In reality, it is a question of the disturbance of animating rhythm. There is no such thing as a rigid conformity with natural law, but only a constantly self-changing interaction triggered off by constant fluctuations in temperature. If these are regulated so as to ensure that a predominance of recreative (refreshing) influences of temperature is maintained, then the enormous problem of more and more people and less and less food will also be solved.

There are two fundamentally different types of motion, both in principle and nature, which respectively initiate different kinds of excitation in bipolar mixtures of basic elements. In this connection, water, the physically first-born and the very blood of the Earth, plays the role of the mediator in both natural and unnatural energetic syntheses. At the same time it is also the recorder of the changes in potential that have already occurred. From this it can be determined what *form of energy* prevails in this *natural accumulator*.

The decisive factor here is that contemporary technology is only aware of and only uses those methods of conveying or moving various media which initiate a reactive increase in pressure and heat. These methods will here be termed 'techno-academic',[64] because they are taught in all academies, universities, high schools and technical colleges. Albert Einstein used them to create a form of energy destructive of all Creation. With this, modern technology has achieved the peak of performance in self-annihilation and turned itself into an absurdity.

From transmissions from ancient times, it is quite clear that those peoples of Indo-Germanic origin or Aryan culture favoured a different form of mass-motion. They overstimulated it, however, and disturbed the unstable state of equilibrium in the other direction. This resulted in whole sections of the Earth being torn upwards (the cataclysm of Atlantis). They strayed from the golden middle way. Existence or non-existence is dependent on the rhythmical dynamics of bipolar forces. We can see this in the alternation of night and day which serves the constant improvement and perfection of the processes of reproduction and higher evolution.

Formative forms of motion are hallmarked by the development of an atomic negative pressure. Professor Ernst Ferdinand Sauerbruch, a surgeon, discovered it in 1908. He realised that without the existence of negative pressure between the lung-surfaces and the pleura, any inhalation and exhalation of diffuse (purified and exalted) precipitates of solar energy (chemically termed oxygen) would be impossible. In other words, there

---

[64] The ancient Greek word *Technao* means *remote from life*, unreal, therefore unnatural and contrary to Nature. – V.S.

would be no re-*spire*-ation (fertilising) of digested, nutritive ethericities, which in exalted form enter the hermetically sealed blood or sap by way of diffusion. Unfortunately Professor Sauerbruch failed to see that this force, which internally bonds (emulsifies) the ethericities of fructigenic and seminal matter, is the *diamagnetism* discovered by Michael Faraday in 1845. Diamagnetism is also known as healing magnetism, animal magnetism or *mesmerism* – an *ism* that science views with disfavour. It manifests itself in a similar fashion to the pressural atomic forces that move a dowsing rod. Like the previously mentioned magnetism, this latter force can neither be measured nor weighed exactly and is hence rejected by science.[65]

There are some forms of energy that are known to alter the values indicated by measuring instruments and on occasion actually to contradict them. Thus it would appear, for example, that the speed of light is constant and therefore invariable. The same also holds true for the so-called acceleration of free fall, because there is no measuring device that can differentiate between *specific* and *absolute* weight. In terms of physical mass, the former is to be construed as breathing (animated), whereas the latter is to be understood as less-breathing, the almost-lifeless.

Before Galileo arrived on the scene with his discoveries, the theories of Aristotle held sway and it was believed that providing there were no disturbing influences, heavier bodies fell faster than lighter ones. That this was actually related to the different fall-velocities of bodies having either specific or absolute weights was completely disregarded. These constantly vary according to the state of their intrinsic qualities. If sufficient qualigen is concentrated in water, for example, then gravity can be partially overcome. Speculative thinkers will be unable to understand this, however, unless they observe Nature's instructive examples very closely.

It must be stated at the outset that in this case the decisive but previously overlooked concentration of qualigen is *only* possible with the aid of radial→axial motion. Every medium becomes heavier if it is unable to breathe and its life is removed. This occurs immediately if excessive influences of heat and pressure are brought to bear by the opposite form of motion. In this case the diamagnetic, or adsuctional and insuctional force (namely Professor Warburg's life-energy) and otherwise known as *levitational force* by people of ancient cultures, is lost.

---

[65] In his book, *The Pattern of the Past* (1969), Guy Underwood writes that: "*Observation of the influence which affects the water diviner suggests that a principle of Nature exists which is unknown to, or unidentified by science. Its main characteristics are that it appears to be generated within the Earth, and to cause wave-motion perpendicular to the Earth's surface; that it has great penetrative power; that it affects the nerve cells of animals; that it forms spiral patterns; and that it is controlled by mathematical laws involving principally the numbers 3 and 7. Until it can be otherwise identified, I shall refer to it as the Earth Force. It could be an unknown principle, but it seems more likely that it is an unrecognised effect of some already-established force, such as magnetism or gravity.*" – Ed.

If the importance of this formative, synthesising and levitating force, produced by centripetal motion *only*, had been recognised for what it is, then the tragic error made by Newton and Leibnitz would have been avoided. They believed that life could be comprehended with great precision through the methodology of mathematics (integral and differential calculus). It is this erroneous reasoning that the whole world has to thank for the evil that Albert Einstein foisted upon us through the exploitation of an all-destroying atomic energy.

A simple experiment demonstrates that a predominantly axially→radially moved (and therefore de-animated) water begins to breathe again and to rise upwards autonomously, when under certain preconditions it is simply made to move planetarily. All forms of growth and more highly-evolved forms of life draw in diffuse (purified and exalted) oxygen. Where unpurified oxygen is impressed from without, disorder and cancer come into being, because cancer is an entity that has been expelled from a state of order.

If pure, nitrogen-free oxygen is *forced* into the lungs of a child over too long a period, then it goes blind. Adults are threatened with an incurable lung inflammation. At high altitudes the more thoroughly the Sun *in-presses* the body with unpurified energies, the sooner blood and sap begin to decompose. The same also applies in forestry, when for the sake of greater quantity and rapidity of rotation, shade-demanding trees, which have far too thin a protective bark, are exposed to direct sunlight, producing unhealthy, 'light-induced' growth. This also happens if forests are excessively thinned. However, if a forest floor that has lain for centuries in protective shade, is warmed above +9°C by excessive clear-felling and thinning, then reafforestation is only possible with deep-rooted, light-demanding species. Such reafforestation is fraught with enormous problems. The young plants have first to grow with difficulty under the scalding light of the Sun until they are large enough to provide the ground with new protective shade. Wholesale karst development (erosion) can only be prevented where the slope is not too steep and the pressure of solar radiation is not excessive.[66]

Almost all monocultivated, plantation forests originate from this ignorance. They can therefore quite properly be described as cancer-prone. They lack diamagnetic, adsuctional and insuctional force and can no longer produce fertile seeds. As is the case with the remaining, over-cleared, old-growth forests, they too can be considered lost if the present methods of forestry are not drastically and quickly altered. Today's fertile soils and whatever water still remains are already over-acidified and polluted to such a degree that there can be no recovery without radical remedial measures.

In Nature everything is reversible. The last hope for restoring the natural order of things lies in the mechanical production of diamagnetism which

---

[66] While not encompassing the full import of the use of the word 'pressure' here, the physical pressure of sunlight = $4 kg/km^2$ (22.8lbs/sq mile) = $0.004 g/m^2$. – Ed.

comes about in the opposite way to the present generation of electricity. The idea of generating life-giving energy mechanically and restoring water's life-force and ability to breathe may sound fantastic. Yet, the behaviour of water itself shows us how this comes about, as it sinks lifelessly back into the womb of the Earth only to rise upwards again as noble water in mountain springs. The reason why it has been impossible to copy this natural process so far is simply because the enabling form of motion was unknown. People mistakenly believe the Earth's rotation to be circular. With circular motion, however, no *negative pressure* or *drop in temperature* can be achieved.

Since the present way of looking at things is founded on a false world-view, humanity has set foot on the road of an unstoppable, regressive proliferation of cancer. As long as the mediative substances of water and air continue to be moved incorrectly with the exclusive use of centrifugence, any production of qualigen or increase and improvement of water and growth will be impossible. Qualigen can only be produced with *cold systems of flow*, in which the more thoroughly atomised the substances (sediments) contained in the water become, the more powerful the life-energies.

Here it should again be stressed that if the media of earth, air and water are moved centrifugally, then reactive, *decentrating* forces evolve from the resultant unnatural intermixture of bipolar concentrates. These also develop if any medium is over-illuminated and/or overwarmed by unpurified (unfiltered) solar radiation. It can therefore be stated that in this case the pressure and heat associated with it increase by the square of the speed of motion. In both cases, however, the resistance to motion also increases – that is, a retroactive force evolves which, viewed ecologically, represents a decomposive energy or its other form, cancer. In this instance matter is rendered lifeless. Oxidising processes follow, and since the levitative forces have been reduced, matter becomes specifically heavier. Red-hot iron, for example, is heavier than cold iron. Einstein was right in his deduction that mass – the concentrate of basic elements – was dependent on velocity. However, it is the *type* of motion that determines whether *decentrating or concentrating* forces arise. In the first case pressure and heat are produced and in the second, a *negative pressure* (vacuum), which *intensifies* by the square of the centripetally-in-winding orbital velocity.

In this way it is therefore possible to transform ordinary air at room-temperature into chemically pure water in a fraction of a second. This is associated with a 1,700-fold[67] decrease in volume, through which an atomic negative pressure evolves, whose attractive force is stronger than

---

[67] Water = $1g/cm^3$. Air = $0.001226g/cm^3$. Water to Air ratio = 1 : 815.66. – Ed.
"At a temperature of 15°C water is 819 times heavier than air at the same temperature. Water-vapour, on the other hand, absorbs a 1,700-fold volume of water. With the evaporation of 1 litre of water, about 600 heat units become latent (stored, bound)." (*Implosion* Magazine No.9, p.26) – W.S.

any explosive pressure.[68] The precipitation of rainwater described above therefore enables the production of rain in any quantity artificially without the formation of thunderstorms. Such water can then be converted into the noblest mountain springwater, provided it is moved planetarily for a short period, as happens in the womb of Mother Earth.

The technologist of today has been taught how to think logically, but not *bio-ecologically*. Therefore it is particularly difficult for him to appreciate the possibilities that ecotechnology offers towards the re-establishment of the proper and natural order of things, which has been disturbed at its most fundamental level. The time has come to free humanity from its dependence on coal and oil by creating those differences in potential that can be converted into kinetic energy. Since coal and oil will be exhausted in the foreseeable future, nuclear fission will be all that remains to banish the spectre of energy-shortages. This is the greatest delusion of all, however, because the radiation and waste accruing from atomic energy will gradually contaminate the air we breathe to such an extent that the mere inhaling of it will be lethal. Humanity therefore stands at an historical crossroad – but in all probability it will continue to develop its present technology and dig its own grave. Then one day, thousands of years hence, the perhaps few remaining survivors will speak of the *technical stupidity* of their forebears and shake their heads over the artefacts of a godforsaken science in whose programme destruction took first place.

Science can still be dethroned, however, through the reversal of present principles. Instead of air pollution, ecotechnical air purification. Instead of the collapse of quality, the rise of quality. Founded as it is on rigid and dogmatic laws, contemporary science must therefore be prohibited by the fastest means possible and thus rendered harmless.

Water – An Enigma whose Solution is as far away as the Stars
From *Implosion* Magazine, No.59.

"We must look into unknown dimensions, into Nature, into that incalculable and imponderable life, whose carrier and mediator – the blood of the Earth that accompanies us steadfastly from the cradle to the grave – is water."

V. S. – 1932. (*Implosion* Magazine, No.103, p.28)

A sophisticated university professor once declared in a public lecture that water was an enigma whose solution lay as far away as the stars. In his

---

[68] Professor Ehrenhaft is supposed to have calculated that on average implosive forces were 127 times more powerful than explosive forces. (*Implosion* Magazine No.83, p.14). – Ed.

view, if it were possible to make water gas-free and to charge it with energy, it would become an explosive capable of blowing up the whole world.

Water can only be rendered gas-free through electrolysis, which decomposes it into hydrogen and oxygen. If a catalyst is introduced, then all that is left is heavy water. On closer examination this discovery, which created a sensation a few years ago, proved to be developmentally harmful, and is therefore worthless culturally and economically.

Expensive experiments with oxyhydrogen gas to assess its suitability as a fuel were equally unproductive. The only remaining option, therefore, is the primitive, but expensive generation of steam, which nevertheless is a reliable source of power. It is known that atmospheric pressure plays a role in the generation of steam. The less the pressure, the more quickly the water boils. Excessive extraction of positively charged air pressure causes water to freeze.

Even these two findings produced no useful effect of any kind. In the first instance the potential force was lacking. In the second, the production of ice was too complicated and too expensive. Even in the raising of water there has been no major advance, apart from purely mechanical improvements to old systems of pumping. The water raised by suction or pressure was none the better for it. On the contrary, it became worse and worse. It was full of gas and stale. This process is also expensive and unprofitable.

The same applies to contemporary river regulation, to the raising and lowering of the water table and to the installation of turbines for generating electricity. The water constantly deteriorates. Ignoring this evidence, all attempts to provide suitable water for agricultural purposes failed.

Compared to all other systems of agricultural fertilisation, the age-old methods of irrigating the land still prove to be the best. However, this is only true to the extent that good water is still available. Today an untouched natural forest is the only supplier of good water. Its quality declines very rapidly, however, if the forest is turned into a monocultural plantation. This calamity was further aggravated when, for purely technical reasons, clear-felling and over-thinning operations became the preferred practice. The drying up of springs and the sinking of the groundwater table were the solid results. Higher-quality timbers disappeared. Mountains streams went beserk. Rivers became silted up through the excessive deposition of sediment, and further sinking of the groundwater table followed.

Science failed totally wherever and however it laid hands on the *ur*-source of evolution and whatever evolved from this carrier and mediator of life. Under such circumstances it is no wonder that water was described as an enigma whose solution lay as far away as the stars, and thus a riddle no human brain could ever fathom. Forced to capitulate, these experts lost all right to describe themselves as competent authorities in this field. Moreover, they also lost the right to adjudicate when the profound secrets of this highest national asset were revealed by a rank outsider.

'Good water – good life! Bad water – bad life! No water – no life!'

This maxim, as simple as it is true, should provide ample food for thought for those who claim to be authorities on water. All they ever do is stabilise the riverbanks first in one way and then another. They take no account of the *water itself*, which obeys the same laws as the blood in our arteries and the sap in plants. The assertion that water should be viewed as the blood of the Earth and the soil that feeds us is completely justified.

At all events it is a curious fact that those most sensitive to criticism are those most in need of instruction. No other profession has failed as miserably as that of the regulators of water and water-resources management. Those who cannot stand criticism already feel guilty. They are well aware their arguments are vulnerable and are concerned only for their own well-being. Unfit for truly public-spirited activities, they are anti-social parasites who should be removed from their positions, and the sooner the better!

The Shattering of Quality

The generation of steam is contrary to Nature. It is unnatural and therefore wrong. In Nature water is not *vaporised*, but *evaporated*. Evaporation should be viewed as a natural process of growth and transformation, of reproduction and higher evolution, and of increasing and raising water quality. The frequency of atmospheric water is higher than that of groundwater. Groundwater first has to be suitably swirled around in order to build up its inner potential energy. This can also be interpreted as the *frequency* of water.

It is Mother-Water's creative capacity that gives birth to the Earth. Rain or precipitation is merely the product of a reduction in potential. It is not a product of *condensation*. It is a manifestation resulting from a concentration of up-flowing energies, which subsequently are downgraded into matter through the influences of heat and light. The cooling of the atmosphere after a fall of rain is therefore the result of a reduction in potential or a loss of internal frequency. Naturalesque evaporation is the prerequisite for the *egenesis* or upward outbirth of the parent-energies of water.

Whatever falls earthwards as rain has failed to achieve its full evolutionary potential. It must return to Mother-Earth to be swirled by her again cycloidally, and thus infused with new levitational energies, before it is able to repeat its earlier attempt to rise. If it succeeds at the second or third attempt, then it is not *water* that comes into being. What evolves is a mysterious pair of potencies to which everything on this Earth that crawls and flies owes its life and movement. Growth, development and increase in genetic diversity (improvement of the species) are dependent on the increase and intensification of these bipolar potencies. The so-called 'climate' is nothing more than the retroactive effect or the inversion of a cultivable evolutive power, which has its beginnings in the Earth. The same applies to natural transformative processes within the Earth itself. This has its own special sphere, the formation of which is associated with very specific preconditions.

Without any inter-exchange between spheres there would be no alternating potential. Nothing in Nature *ever* happens directly or in any logical sequence. Every transformation results in a complete change in the type of inner potential. This is why mechanistic river regulation is inevitably doomed to failure. It is *rhythm* that orders transformation and it is only in this way that the '*transmutable*' (water) can retain its natural properties. It will become the most dangerous enemy of every living thing, however, if it is no longer able peacefully to pursue its way of life and to regenerate its power of reproduction and upward evolution. In the light of this, the present state of the world and *all* contemporary world-views, including all existing doctrines, axioms and dogmas, will have to be changed equally fundamentally.

Even air-pressure is subjected to constant change. The higher the altitude, we are told, the more rarefied the air becomes. At the very highest levels there is supposed to be an absolute void, a no-thingness. This is where fundamental error and nonsense, and the dangerous influences of religious belief begin. These would be viewed quite differently today, had the natural processes of evolution been allowed to continue. This belief would then be just as impossible as belief in the Law of Constant Conservation of Energy and the absolute omnipotence of God. Without the build-up of potencies, without the increase of potentiated matter and without an intensification of these potentialities, even Lord God would be powerless.

All these things are entirely dependent on the cycloidal movement of water and air. Nothing can reproduce and upwardly evolve itself of itself, nor can it increase and qualitatively improve itself exclusively with its own internal resources. We are therefore confronted by the greatest crime and stupidity of all time: the elimination of the *cycloidal movement of water*. This has buried alive the source of all synthesis of quality.

Had a little more attention been paid to the inner behaviour of water after the mechanical extraction of its gas-content, which immediately changes the freezing and boiling points, the conclusion could then have been drawn that opposites react differently to suction and pressure. The same is true for the changes that occur when water is boiled and subsequently re-cooled, which also alters its freezing and boiling points. It should also have been possible to understand why water cannot be raised higher than 10m (33ft) with suction pumps. If water is moved naturalesquely, however, then it rises to any desired height almost without any mechanical inducement. Consequently all contemporary systems of pumping lose their validity, which in any case are very expensive to operate, quite apart from degrading the water itself.

Naturally-moving water multiplies itself. It raises its own quality and wells up autonomously. It changes its freezing and boiling points. Wise Nature makes use of this phenomenon to raise water to the highest mountain peaks without pumps, as in the case of high mountain springs. The term 'raise' should not be taken literally, since here it concerns a natural process of reproduction and higher development. This further serves the egenesis of the air, the upward outbirth or creation of an aeriform envelope, which in turn enables the evolution of higher forms of life.

When the increased, ennobled, specifically heavier, dense and cool water springs forth from a spring, it surrenders its surplus levitational substances. In consequence it charges itself with gravitational substances and flows downwards. Starting from the very smallest beginnings, it gradually develops into a mighty river.

If a spring is divested of its natural protection (shade-giving trees, cool rocks, etc.) then the cool spring-water becomes warm and absolutely (not specifically) lighter. Shortly thereafter it loses its natural power to rise and in consequence disappears. *No water – no life. Here springs – there life.*

The Energetic Phenomenon above the Zone of Air

What lies above the atmospheric envelope is neither rarefied air, nor is it a vacuum. It is a phenomenon of *negative energy*, the opposite of positive air-pressure. This very special kind of low or negative pressure comes into being when geospheric, levitational matter atomically consumes (binds) substances, creating positive air-pressure. Thereupon what Goethe called the 'Eternally Female', comes into being. It draws everything upwards and to itself and also maintains the heavy Earth-masses in unstable equilibrium.

The tremendous weight of the Earth-masses, therefore, has the task of restraining the levitational forces. The weight of these masses alone would not be enough if the physically-ponderous air-pressure, or more accurately, the most finely-dispersed varieties and masses of atmospheric water bearing down on the Earth, did not reinforce the Earth's weight. The Earth's force of *gravity*, therefore, is insufficient to prevent the Earth-masses from floating upwards. In order to overcome this so-called gravity, all that is necessary is to bind the over-burdening air pressure atomically, in the way that wise Nature shows us in every vortex of air. Then every effort has to be made to brake this upward force.

The art of braking will not be elaborated in any greater detail here. We are more concerned with generating impulsive motion naturalesquely, so that carrier-substances of the air can be built up. In other words, the *propellent* must first be generated in the atmosphere, which geospheric surpluses can then consume. The outcome of this is the true levitational force that holds the Cosmos together. The almost cost-free possibility of overcoming the so-called force of gravity depends upon the naturalesque binding and generation of the essential propellent.

Without gravitation there would be no levitation, and conversely, no down-fall without a prior up-rise. And now to the question: which of these is more important or takes precedence? The answer is not all that simple, because both processes have to take place simultaneously. The egg, so to speak, has to be born at the same time as the hen. In this context Goethe writes: *"Nature is neither core nor covering; she is both at once"*. In practical terms this means that there are fructigens, fertilising substances and carrier substances. These are distinct opposites and move in different directions. However, all of them have to operate on a common axis, whereby fertilising substances have to be moved by centrifugence, and fructigens by centripetence. In addition they must all rotate about each other.

This art is called the *cycloid form of motion* – where care must be taken to ensure that the fructigens move at a speed corresponding to the sum of both speeds squared. This speed is necessary and central to the aim of building up *qualigen*, from where the Goddess 'quality', the immortalised 'Eternally Female', the marvel of levitation, is ultimately born. This involves the rebirth of a mysterious *ur*-force, which comes into being when the power-supplying medium (oxygen) pressing down on the Earth is consumed in the realm of the atmosphere, or becomes bound in an atomic (diffuse) state.

World Domination through the Destruction of Quality

If humanity is to be controlled, quality has to be destroyed. If the excellence of water and the wholesomeness of food is destroyed, then the quality of human thought-processes deteriorates. The simplest way of achieving this was to eliminate the cycloid movement of water. Science was established and launched on its misguided path in order to develop a technology and mechanical power that exploited a form of energy which, in Nature, leads to degradation and destruction.

It is hardly surprising, therefore, that under such circumstances few or no elements of quality could evolve. This resulted in a spiritual and mental torpor, the prerequisite for mastery over the broad mass of the people. Science invented the Law of Constant Conservation of Energy and in addition forbade any questioning of it. Against this background of increasing intellectual apathy, science then undertook the deplorable task of *"leading all mankind around by the nose on an ever more barren heath"*, as Goethe so aptly put it. For lack of quality and substance the struggle for survival became more and more difficult and beyond all endurance. Was it thus such a surprise that the poor wretches of humanity had to be promised eternal salvation in another realm?

A poet once wrote:

Into thy daily occupation,
Transience needs interpolation.
Yearly thou'llst have confirmation,
Nought's more true than transformation.

Naturally it requires a certain courage to tell the truth! There would, however, be no point, if it did not arouse curiosity as to how the 'transmutable' water could be produced with the aid of a suitable device. The only way these can be built is by emulating Nature's processes. Mechanical power will then be free as no intelligent person will still continue to build an explosion motor. It is equally obvious that once the unholy trinity of business, science and religion have disappeared without trace, all humanity will once more awaken to a life worthy of human beings.

This can be achieved by such machines as the *Repulsator* and the *Repulsine*.[69] The former is used to create a superabundance of food, and the latter to create the almost absolute and continually renewable freedom of movement. How the Repulsator can liberate the world from hunger will now be addressed.

The Repulsator and the Superabundance of Food

The Repulsator is a faithful copy of the Earth and its functions. In order to be moved cycloidally, warm and stale water sinks into the Earth where it is subjected to the influence of geospheric energy. In binding this positively charged propellent the water becomes negatively charged, reconstitutes itself, and bursts of energy stream upwards to great heights. Some of these will be absorbed by the roots of vegetation and transformed into matter. The remainder rise into the atmosphere, there to be forced to interact with cosmic energy.

In the Repulsator the same process of growth and transformation takes place. However, the products of revaluation (non-spacial, formless, structuring forces) are not able to escape, because the containing vessel is well insulated and sealed. In this way unlimited quantities of levitating

---

[69] These two devices are described in *The Energy Revolution*, vol.4 of the Ecotechnology Series. – Ed.

mother-elements can be successfully incorporated into stale water. All reserves of worn-out water can also be turned inside out and disintegrated for the purposes of forming new water. The finished water can barely be differentiated from high-grade springwater. If such water is sipped slowly, then an impotent man will regain his potency. This water restores youth and rebuilds uplifting energies. It gives the human, animal and plant organism renewed potency, life-force and mobility.

Paracelsus called this elixir of life, the *ur*-stuff of love. He was unable to discover it himself because he used fire when trying to prepare it. Leonardo da Vinci named this primary energiser of life '*Il Primo Motore*' (the prime mover). Its distillation tormented him all his life. Goethe referred to this uplifting substance as the 'Eternally Female'. More precisely, it should be called the 'Immortalised Feminine Aspect', the growth-promoting Goddess of Quality. It is She that enfolds the Earth and maintains it in a gyrating, unstable state of equilibrium, and it is only through cycloidal motion that quantity can be transformed into levitational or uplifting force.

In Nature this process takes place continuously which, expressed as life-energy, is the reason for the progressive intensification and increase of true *ur*-potency. Coming from the Beyond, it is a higher influence that activates growth. Growth is none other than the waste-product of qualigenic energies, which as they rise upwards are solidified by the direct influences of light and heat. This also explains the extraordinarily vigorous growth when plants are watered with such mother-water. The true process of noble fertilisation takes place according to a principle similar to the one Sven Hedin describes in his book *The Flight of the Great Horse*.[70] Called 'Holy Water' by the ancients (a variety being genuine Greek fire-water[71]), this extremely potent mother-water is incombustible. It does ignite in the light of the Sun, however, if it is overcharged with phosphorus compounds.

If such water is atomised in a cylinder and a modicum of atmospheric oxygen is introduced, then all that is needed is the slight heating pressure of a descending piston to transform, in a split second, the highly potent mother-water into an aeriform state. What happens is that an instantaneous egenesis (upward outbirth) of air occurs, leading to a 1,700-fold increase in volume. This potential pressure can be intensified at will by repeating the process, because the decisive ethericities are almost non-spacial and without form. Therefore they can be incorporated into this natural dynagen accumulator in unlimited quantity, the accumulator itself becoming a reservoir of

---

[70] See references in 'Organic Syntheses', p. 153 of this book, and 'Natural Farm Husbandry' in *The Fertile Earth*, vol. 3 of the Ecotechnology series. – Ed.

[71] 'Greek fire': Is a combustible, highly ignitable composition, which caught alight when wetted, for setting fire to an enemy's ships, also known as St Elmo's fire, *elixir vitae* or corposant fire. (O E D & C E D). – Ed.

positive air-pressure. Having been impregnated by heat pressure, the highly potent mother-water builds up an enormous pressure of air. In the initial stages of transformation the capacity of this gas-less, explosive water, charged with negative products of synthesis, is in the order of 2,000 atmospheres. With the addition of the appropriate trace-elements and a longer period of circulation during the developmental phase, the pressure can be intensified to any desired level.

The exhaust gases immediately try to relocate to a higher level and exert suction on an appropriately designed piston-head, which reinforces the expansive pressure in the cylinder. These uplifting substances then escape if the piston-stroke is of suitable length, whereupon the piston can descend almost without friction. At this juncture a more detailed description of this ideal expansion motor may not be given.[72] Whoever is the first to possess this expansion motor, which operates almost silently and without cost, has victory in his pocket. Submarines in particular would have an unlimited operating range, because fuel is abundantly available and the exhaust fumes are the best mountain air.

---

[72] This was due to then current patent applications and Schauberger's need to protect his inventions from abuse and suppression – Ed.

On Energy, Eggs and Natural Motion

Our Motion Is Wrong
Letter from Viktor Schauberger, 1954. (*Implosion* Magazine, No.19)

Dear Dr M G,

I thank you for your interesting letter and I would like to answer your questions in a very unscientific way thus. The old foresters, huntsmen and fishermen and their wives at the spinning wheel, sought through old fairy-tales, to explain the mysterious steadfastness of trout standing motionless in the axis of flowing springwater.

It was rumoured that the souls of the departed were to be found in such springs. They were supposed to be slowly released from their decomposing earthly remains in the cool womb of Mother Earth. Gradually coming together in tumbling streams and migrating upstream, where upon reaching the spring they became charged with a purifying ascending current. This current also draws a certain quantity of springwater with it in order then to be able to rise heavenwards.

As a four-year old, I wanted to have a close look at this migration of souls and in the process fell into the ice-cold water. The maid-servant fished me out and administered stout blows, shaking the water out of my lungs and stomach. She then took me into the kitchen and angrily sat me down on the kitchen sideboard, where my shocked mother took over. While my clothes were being changed, she delivered an unforgettable lecture.

*"You silly boy! How on earth can you go to **that** water, where the poor souls journey towards the mountains. They will attract you, pull you in, drown you, and then you die and have to go with them. When you are grown up and not before, if one day have pressing problems, then, my young man, go to the springbrook, where my soul will also be and which will then give you motherly advice and help you when I am no longer on this Earth."*

Thirty years later as a young forester, I was going to lose my job due to the inexploitability of the timber in remote stands of a forest reserve. It was due to be administered by a cheaper gamekeeper instead. Standing near a high-lying crystal-clear stream I gazed ruminatively into the fast-flowing water, which rose from a crack in the rocks a few hundred metres higher up, and murmured into it. My mother's spirit did not answer and, disappointed, I wanted to cross over the stream, using my staff as a vaulting pole. Trying to find a secure purchase for the end of my staff on the smooth, rocky stream-bed, I flushed a stationary trout from its lair, which fled upstream like lightning.

Two questions shot into my head: firstly, how do trout reach these high places? Secondly, just how was it possible that there, right in the axis of the current, a large number of fish were able to stand almost motionless? Steering themselves with only slight movements of their tail-fins, how were they able so effortlessly to overcome their own weight and the specific weight of the heavy water flowing against them?

Was it the souls migrating towards the mountains that drew the fish along in their wake? Or did an axial, biomagnetic force prevail which also prevented the downstream acceleration of heavy logs, and yet which, through mysterious counter-forces of suction, made their transport down the central axis possible? To cut a long story short, this is how I discovered animalistic earth-sap-blood magnetism, which enables the naturally flowing, planetarily inward-spiralling water-masses to maintain their steadiness of flow in variable gradients. This steadying force is rendered inoperative if the watercourse is regulated and straightened out. It is also extinguished if springwater is centrifuged in high-speed, iron (steel) pressure-turbines.

Professor Ernst Ferdinand Sauerbruch discovered the organic low-pressure chamber that lies between the pleura and the surface of the lungs, which functions as a biological vacuum and enables the inhalation of breathable matter and the resistance-free expansion of the lungs. Professor Sauerbruch was not aware, however, that this phenomenon is only possible through planetary motion. In this process positive atmospheric pressure is countered by a negative pressure, variable at will, whose power increases by the square of the in-winding rotational velocity. Moreover, if water is centrifuged, positive pressure increases in precisely the same measure, resulting in the lateral ejection of bipolar sediments. The naturalesque way, however, is to centripetalise them along the central axis. Otherwise a bio-electric nuclear axis surrounded by a decadent magnetic field will develop (as with electromagnets) at the same time as a diamagnetic, oxygen-repelling, but nonetheless iron-attracting magnetic function. In this case a pole-reversal occurs, i.e. an atomic excess pressure evolves in the longitudinal axis of a watercourse, which makes all respiration and insuction of diffuse, atmospheric oxygen impossible. The water inevitably suffocates, and with further direct or indirect over-illumination it decomposes due to the increasingly aggressive concentration of oxygen.

If water is made to move in a predominantly centrifugal manner, an X-ray-like radiation is produced, which is the result of warm processes of distillation (fermentation) in the zone where the raw materials and basic substances are present. This mainly emanates in a horizontal direction, drawing decomposing matter into its wake and encrusting the wall-surfaces, as occurs in sclerosis. It triggers off *additional* heating effects which naturally are at the expense of formative and levitative atomic energies. This occurs when the temperature of a river is raised to bathing temperature in summer, which, if the rate of flow approximates 800 million cubic metres (1,046 million cu yds) of water per second, requires a further expenditure of energy of about 60 million horsepower (to heat up $1m^3$ of water by only 0.1°C, requires an additional heat input in the order of 42,700 kilogram metres).

The fact that energies are also bipolar has therefore been overlooked; which form of energy is activated is solely dependent on the type of motion of the medium in question (be it earth, air or water). Through the agency of warm or cold fermentation processes in the raw material zone, it is either the bioelectric, *decomposive* or biomagnetic, *levitative* energy-form that in the end prevails. In an *unnaturally-regulated* waterway or a river channelled through *steel* pressure turbines, increases in heat occur, which are identical to those in atomic piles when uranium (which is actually represented in every cell as a form of sediment) is bombarded with neutrons.

Every leaf, every pine-needle, every eggshell, gill, lung, or piece of bark is a diffusion system that possesses the same accommodating capacity as the pupil of the eye. It dilates or constricts, permitting only the passage of regenerative, refreshing and almost heatless oxygenic elements. It also absorbs those elements that are developmentally harmful. More precisely, it ejects them centrifugally, because they are endowed with enormous velocities. An indirect centrifugal effect is activated in this way, which purifies the entraining, half sucked-in, half pressed-in seminal matter. Any motion of bipolar entities always evolves through the agency of two components of motion – and what really matters here is which of the two components, suction or pressure, predominates. The regulation of this rhythmical (intrinsically dynamic) interplay of forces so that the suction component prevails is an art, which is only possible through a predominantly planetary form of mass-movement.

All over-illumination or over-heating, whether this takes place directly or techno-academically, *weakens* the biomagnetism that attracts and draws in animalistic oxygen. The same also occurs when toxic fumes are inhaled. Asthma and blood circulation disorders are the biological consequences of the type of motion that intensifies heat and reverses the polarity of the biomagnetic, axial force.

I could write you a book showing the errors committed by those in academies, universities and schools of higher education, who only know how to move bipolar masses when centrifugence predominates and *pressure* and heat intensify. They are ignorant of the life-curves which are made possible through a predominantly *centripetal*, planetary system of moving earth, blood, water and sap that reduces heat and pressure.

With the latter, giving due attention to the number of revolutions, it is possible to approach the anomaly point; the temperature zero-point of every manifestation of life. At this point the negatively supercharged inner elements of a seed, sediment or a cell become highly active, whereas positively supercharged energy-concentrates become passive. They become condensed, and after being atomised mechanically, are bound by negatively supercharged energy-concentrates.

The end-product of this interaction is biomagnetism. It emanates its surplus energy mainly in a vertical direction. If conducted through vacuum tubes, the tubes radiate a bluish-green shimmer of light. In contrast, the other forces exhibit a dark red, strongly pulsating light-effect at the periphery of a vacuum bulb. The former are the spiritual (atomic), biomagnetic and animating ethericities, which on their upward path attract and draw in surrounding atmospheric oxygen in its diffused state. In the process a tornado-like, upwardly-widening and water-precipitating *ur*-cyclone is triggered off, which draws up the whole mass of the Earth. This force is encircled by an expanding development of solidifying entities, which prevent the rupture of the Earth. The peoples of ancient cultures, who made excessive use of planetary motion, must therefore inadvertently have brought about the break-up of Atlantis.

However, gravitation and levitation can be regulated by mechanically-produced motion. With this comes liberation, independence from physical and spiritual ponderousness. Gravitation really only reigns in our atmospheric living space, for above this weightlessness prevails. Using this insight, I built the *Repulsine* in Mauthausen concentration camp. At the end of the war the models, prototypes and working drawings fell into Russian hands, and as so-called 'flying saucers' they will soon flit around the world like phantoms. These craft do not stem from higher beings, but rather from a forester who merely turned the Ressel pressure-screw inside out. In this way I managed to construct the biomagnetism-generating suction-screw or, expressed more scientifically, the logarithmical tractor-screw. Dr Putlitz calculated it for me and, as an inhabitant of Hamburg in Germany and a former communist leader, he is not entirely without repute. The rest of those assigned to the project in Mauthausen concentration camp were all either Poles or Czechs. I received reports from them after the war that further development of the

Repulsine was being diligently pursued, so any further commentary is superfluous.

The gyrating motion detected by water-diviners is predominantly bioelectric in nature and function. In the majority of cases it is indicative of an intermixture of the diffuse substances in different types of water, where the latter mutually intersect. At their point of intersection an atomic form of rotational motion becomes free and active, which can also be achieved with the aid of dynamos, which centrifuge the sediment of the air.

The same applies to pressure-turbines used in hydroelectric power generation and hence with its *soul removed*, the groundwater must sink over wider and wider areas as the rate of rotation increases through the steepening of the intake gradient.

I have already stated that lower forms of radiation emitted in bundled form by techno-academically moved water ignite fluids and gases. It has likewise become completely obvious that their activity also triggers off inflammation of the blood and sap, called 'cold fire' according to folklore, and correctly named *cancer*. In this regard our miseducated scientists can be described as the true instigators of cancer.

Half of the twenty-six patents I have applied for have already been granted. Their illustration and explanation is correct. They are concerned with a three-dimensional form of motion and the atomic counter-flowing motion arising from it. In terms of its function, it is the primeval breathing force, which attracts and draws into itself the elements of oxygen and earth-magnetism. It is the generator of levitation force.

The will of the father builds houses for his children. Placing a curse on the mother tears them down again. As the ancients foretold, whosoever should disturb the spiritual journey of the *eternally female*, which lifts everything upwards and endows it with the ability to move autonomously, will be forever cursed.

Please read *Stern*, folio 29, of 18th July 1954, which states in this regard: "*We stand before a catastrophe of unimaginable proportions. The world has too little water. America reports new droughts. Denmark has introduced water restrictions. France is closing down businesses. In Germany in particular, scarcity of water is increasing despite alarming rainfall. Wells are drying up. Water supplies are rationed. Factories in the Ruhr are cancelling shifts. Now the artificial production of rain is being attempted by making 'water out of water'. Seawater is to be transformed into poisonous distilled water by processes of warm distillation.*"

Every fifth German is dying from over-acidified food, water and air for breathing. It is the origin of regressive development and of the onset and cause of cancer.

Those who refuse to listen to advice cannot be helped. Every square kilometre of mixed forest soil retains about 120 million litres of rainwater. Countless diffusion systems abstract oxygen from it and conduct this most finely atomised rainwater into the ground, where with the aid of planetary Earth-motion, it is once again transformed into levitative spring-water.

The Danube has a catchment area of about $90,000km^2$ (34,750sq m). $46,000km^2$ (17,760sq m) of this lie in Bavaria. If only one sixth of the erstwhile forest had been left standing, then about 180 million m^3 (235 million cu yds) of water would have been retained, and the recent enormously destructive inundations would have been totally impossible. As a result, Germany must now spend 10,000m Marks (see report in *Der Spiegel*, 32, 1954) merely to alleviate the grossest water damage. No scientist knows how it could have happened. Were it known, then the madness and stupidity of de-spiritualising or de-magnetising the blood of the Earth through techno-academic forms of motion would not have occurred.

Everyone feels assailed when one tells the truth. The dearth of water will therefore inevitably result in an economic collapse. People could not have gone about it more idiotically and in so doing bring about their own demise. Either our present technical systems of motion will be prohibited by law or all humanity will be lost in zones of greater heat. Our children are confronted with a grisly future. They will have to dig for good water in the same way as today we dig for gold or other supposed earth-treasures. A totally stubborn scientist once told me curtly, *"Few people ever listen to the bleat of the camels[73] of the advancing Bedouins"*.

The camels already arrived long ago and their bleating only accelerates the further decline. I regularly ask myself; will those who have caused this global calamity beat on their heroes' breasts and smash themselves to pieces? Will they bawl *"mea culpa, mea culpa!"* and accuse themselves? On the contrary, they will leave no stone unturned to save their prestige and in the process fail to see that they will be the first to be hung, once despair throughout Europe causes an abrupt public about-face.

With atom bombs we can never extricate ourselves from this mess. The evil genies have been summoned and will not be laid to rest. Through the decomposive effect of radioactive radiation, even atmospheric water will now be electrolysed. Droughts here, flood catastrophes there at ever shorter intervals. Whatever degenerates becomes unstable.

Dear Dr G, the situation is hopeless, and I am happy that I am already so old. The wearers of robes and gowns have engineered a gruesome state of

[73] In German the word 'camel' can also refer colloquially to an 'idiot', therefore in the following sentence this other interpretation should be borne in mind. – Ed.

affairs and will continue to foster this work of the Devil unfalteringly. Such is my opinion,

Yours sincerely, Viktor Schauberger. (1954)*

The Coming Bio-Ecotechnical Age
From *TAU*, No.146 (June 1936)

For decades I have been of the opinion that it is impossible to continue long-term to exploit mineral reserves of coal, reservoirs of oil and hydro-power as they are exploited today. These energy-concentrates harbour the raw elements for the build-up in foodstuffs and we human beings must perish if we cut off the very branch we are sitting on. My researches into the nature of water have unearthed such a plethora of quite bewildering insights that we simply cannot so thoughtlessly continue to let slip the possibilities offered here.

I have discovered the *Mutator*, which enables contemporary methods of power generation to be eliminated altogether. The best fuel is air which, when decomposed organically in capillary tubes (double-spiral pipes), enables the indirect metamorphosis of spacial substances (gases) into non-spacial matter (electrozoic energies). Through this rearrangement of atoms (atomic transformation – not atomic disintegration) it is possible to generate heat and cold organically in the same way that Nature shows us every-where. As a further consequence, the possibility presents itself not only of producing the other transitional energies of *light* and *dark* artificially, but also the *ur*-produced intermediate compounds in Nature, which we commonly describe as proteins. These substances are produced through concentration of the three spacial dimensions of solid, liquid and gaseous + *cosmic radiation*. The result of these five-dimensional interactive influences is:

Juvenile earth – juvenile water – juvenile air.

These natural basic building blocks carry within them the impulse for further transformation and development (growth). Once we have researched the fundamental laws of growth, then the primary pre-condition for the emergence of life is provided, because one is then in a

* This was written four years before Viktor Schauberger died. At the time of writing the world was preoccupied with reconstruction after the War years, which eventually led to the excesses of the consumer society and the naked desire for acquisition of material wealth. With so much war damage to be rectified, the damage to Nature and the environment seemed at the time to be of little consequence. The technological successes arising from war research had also given a huge impetus to the advance of an unnatural science and technology at the expense of Nature. With this as a background it was small wonder that Viktor Schauberger was depressed at developments. – Ed.

position to administer and organise an overabundance at will. *Growth signifies the overcoming of earthly gravity*, the reincarnation and resurrection of the physical remains of former life. The form and function of these remains are reversed through their transformation in the Earth and become the dynamic substances in the immediately following life-phase, which re-animate those still at a lower state of organisation. In this eternal flow, in this uninterrupted process of *form-generating motion and motion-generating form*,[74] one organ plays a decisive role, which we can observe everywhere. This organ is the pulsating double-spiral tube (whorl-pipe), the *capillary* (or the capillary system), which is to be understood as something entirely different from contemporary science's conception of a capillary tube.

Air and water are intermediate organisms whose task is to connect the above and below. These intermediate substances are always enclosed within a third, neutral substance and as 'opposites' move themselves in different ways and directions. To enable this co-active motion a mediator is therefore required, which has two different axes in one and the same profile (egg). Out of this inversely dimensional (morphological) motion, longitudinal potential (magnetism) and latitudinal potential (electricity) develop. As mechanical and physical opposites, pressure and suction are functional dynamic phenomena, which necessarily encounter each other rhythmically and continually through which, according to their proportion and organisation, the individual pattern of motion-generating form or form-generating motion arises. Every manifestation of life of whatever kind is always a chemico-physical or an electromagnetic process triggered by the cathode and anode system of Sun and Moon. If we are able to trigger this impulse in the proper way, then further development occurs automatically and we are thus in a position not only to engender growth of a purely physical nature at will, but also to potentiate (cause the growth of) pure incorporeal essences (kinetic energies). Accordingly we are already thoroughly embroiled in an entirely new technology which only exploits peak energies and which has virtually nothing to do with the over-exploitation of Nature. This new technology leads to humanity's reconciliation with Nature. In future it will safeguard humanity from any further blunders caused by our present technology, which operates on a purely

---

[74] The German expressions here are 'Gestaltungsbewegung' and 'Bewegungsgestaltung', which encompass the German concept of 'Gestalt' or 'Gestaltung', an expression that is untranslatable with a single English word. *Collins English Dictionary* defines 'Gestalt' as "a pattern or structure possessing qualities as a whole that cannot be described merely as a sum of its parts." As such it can also apply to physical, energetic and spiritual forms, matrices and structures. To simplify understanding, the word 'form' has been chosen to represent it. In *The Water Wizard*, vol.1 of the Ecotechnology series, prior to their revised interpretation, these two concepts are expressed as 'motion-of-creation' (form-generating motion) and 'creation-of-motion' (motion-generating form). – Ed.

materialistic basis and lacks the fundamental principles of natural philosophy.

A *priori* this new form of technology demands not only a renunciation of all contemporary ways of thinking in confessional domains, but also a total renunciation of sophisticated thought processes. It requires a profound perception and study of the unified interdependencies in Nature through which what we call the *Universe*, the *All*, the *Whole* is created. Life is not an end in itself, but only a means to an end, a *Fata Morgana* in thousandfold shape and form. It is always the same preconditions that give rise to any given manifestation of life.

Brother stone, brother plant: these are words of profound meaning and significance! The great art is to organise and apportion the wonderful inter-relationships in such a way that practical means to an end will result. Call them machines if you will. They are devices, however, which do not seek to overburden Nature, but to support her. Their purpose is to accomplish all that we desire to achieve today without making use of our entirely perverted and contrary technology. Through today's extremes the road towards the golden middle way is slowly unfolding. Those who can take a broad hint and are imbued with will to turn their backs on today's violent processes will become the leaders of tomorrow.

At one point I thought that as an Aryan nation we would be in the best position to give a lead. In every respect my discoveries bear the hallmarks of the knowledge of bio-ecological formation possessed by ancient Asian cultures, because they already knew these secrets many thousands of years ago and possessed silent aircraft (*vimanas*), which were able to overcome gravity by a very simple process. Unfortunately, however, contemporary Germans appear to cling to their technical delusions and do not seem to want to relinquish them. It is therefore to be feared that they will miss the opportunity, because even now people of other tongues, who are already aware of this inner degeneration in Nature, are more flexible and are becoming increasingly interested. Here too we are concerned with *impulses* which evolve at lightning speed once the thought ignites. This has happened, because today it is no longer a question of fantasies, but of facts. These are anchored in causes, which were relegated to secondary importance while spurious concepts were pursued, whose relevance is everywhere beginning to pale into insignificance.

We therefore stand on the eve of the ecological age or we are perhaps already in the middle of it. Biology is materialised idealism or perhaps even idealised materialism. And thus it is that a field of high tension exists between these two conflicting concepts, which someday perhaps will somewhere produce a synthesis.

Movement and Forms

From Special Edition of *Mensch und Technik*, Vol. 2, 1993, Section 3.1.

It has become clear to some that the origins of all manifestations of life and motion lie in chemo-physical and electromagnetic causes, which on their part are actuated by plus and minus temperatures.

Will and counter-will are functions of the temperatures that arise for one reason or another. Work is a function of the temperatures that lie above or below the turning point or point of intersection – the anomaly point – in which all life arises (zero-point). It is of similar nature to what we experience as the 'present', which is a function of an on-going, extremely high velocity flow. This flow is an eternal *form-generating motion*, which in turn gives rise to the continuous *motion-generating form*. Thus we arrive at both the fallacious concepts and the explanation of the life that lies between birth and burial, which is the illusion that clothes evolution with reality. Temperature is thus the difference between differences, out of which the ceaseless movement of evolution arises. This in turn is the product of tensions resulting from the contrasting directions of movement.

The movement of the planets is mirrored in the movement of earthly bodies and so the possibility also exists to order the course of planetary systems by means of a particular physical motion. Conversely, it is also possible to exploit planetary motion to produce physical motion, which will enable us to harness a constantly waxing motive force for our own use. Opposites have their appointed directions of movement and find their expression in the interchangeability of the substances that live and move.

If we can now succeed in dosing like-directional dynagens and in bringing the groups of substances thus organised into mutually opposing motion, then a maximal motion will result. Its harmonic counterpart is the creation of the minimal form of motion, because the latter represents the totality in a single point. This is the turning or anomaly point out of which is born the physically self-embodying *form-generating motion* arising from the *motion-generating form*.

This physical formation is the product of organic formative processes and it is obvious that in order to construct such physical forms, we must make use of certain basic shapes.

This basic shape we find in the ellipse, which once set in motion, produces the mirror-image, opposite form. As the natural counterpart, the latter also creates the opposite temperament or reciprocal temperatures, which on their part give rise to the tensions and the form of motion associated with them.

Since we are concerned here with pure morphological patterns, there can be no state of equilibrium and therefore that state cannot exist, which we understand as 'rest'. In reality this apparent rest is the very highest state of motion and at the same time it is the point of material transformation. It is also the point of *ur*-generation of the most purely spiritual dimension, which transforms itself at the speed of lightning into movement or form, through which the thing in question functions independently and is thus energised from within.

Conversely, this two-fold motion is also the origin of life, which on its part continually transforms itself both without and within.

If we should now succeed in maintaining any given substance, for instance water, in a constant state of change and transformation, then we are presented with the eternal growth-generating motion. This produces the *motion-generating form*, which we can transform by indirect means into that which is described as 'energy' in today's terminology.

With this we are well on the way to a technology that will usher in and order the coming biological age.

The basic principle of this natural technology is the resistanceless motion that naturally and necessarily results from the evolutionary *motion-generating form*. As this form evolves to a higher state of organisation, it automatically vacates the place it occupied as it moves upward as *form-generating motion*.

Bio-Ecological Technology
From the Special Edition of *Mensch und Technik* Vol. 2, 1993, section 3.3.

The picture of the world and its motion, sketched here in broad outline, are the foundations for the future which is biological technology. Essentially this is founded on the possibility of shifting the subjective centre of gravity. Through inner-atomic restructuring with the aid of mutators, the centre of gravity within an inversely dimensional plane begins to wander within and about itself, i.e. it always manifests itself at different points as a subjective present. The precondition for this self-moving, spiralling motion is the planetary construction of elliptical bodies of certain interrelated magnitudes. Their contrasting directions of motion give rise to two inversely proportional energy-fields of opposite potential. This results in the formation of a nucleus that rotates about its own axis and which, like the yolk of an egg, floats freely in the organic centre-point, in the life-point or anomaly point.

This organic anomaly point is the zero-point, as it were, of the elliptical formative body, or the point of rotation of the mutually inversely-moving organic and inorganic masses. This organic motion is the origin of the eternal creation-of-motion and the cause of the ceaselessly changing pattern of day and night, heat and cold, life and death. Every substance of whatever kind has come into being naturally. Hence it is of divine nature, immortal and thus in constantly transmuting form. Whatever appears as spacially visible bodies in the present, reconstituted itself in the past and returns in exalted form as transformed future. Between the purely spiritual and purely physical dimensions of transformation lie the metamorphic and non-spacial intermediate dimensions, which are instrumental in the processes of transposition.

Heat and cold, light and dark are likewise intermediate dimensions of these perpetually self-reconstituting magnitudes. For this reason and with the aid of these wonderful intermediate forms, we human beings are also in a position to move masses with unlimited power. Providing that is, we understand how to organise the special interrelationships harmonically.

All motion is a function of so-called temperatures, which begin to structure themselves, if we place the thing in the organic life-point (the anomaly point of water is +4°C). For this reason the only true and proper work of god-like and thus creative humanity is the organisation of the planetary drive-mechanism. This we must be able to order spacially and energetically (non-spacially) in such a way that an autonomous, organised cycle comes into being through which a constantly self-evolving motion arises, which as so-called 'power' we can then make use of.

This type of use is no ruinous exploitation, but rather the reinforcement of the all-prevailing and moving will (of God). To serve Nature means to command her wisely, and for this reason those who humbly revere Nature, will perceive her mysterious processes and in response will give substance to the will of that which moves and animates everything.

The Age-Old Secret of the Atom
From *Implosion* Magazine, No.74

The physics of today will collapse like an uninteresting house of cards once the age-old secret of the atom has been revealed. Then it will at last be realised that apart from life-destroying atomic energies, which kill off all growth and other forms of life, there are also life-giving energies with a multiplying and ennobling function. Something of their nature will be revealed here for the first time. It must also be stated that in all naturally healthy rivers or in every spring there are particular minerals that are far more interesting than uranium or other energy-concentrates used for purposes of annihilation.

Without higher-grade atomic energies there is no life, and without life, there are no higher-grade atomic energies. In Nature all life springs from so-called transitional bacteriophagous essences.[75] These should be regarded as the *ur*-origin of all physical existence, and more importantly as the source of the physically first-born – water. Life is recitative – that is to say, all deceased life reverts to the metaphysical. Forces therefore also exist whose function is to mortify, and which come into being after the animating forces have been extinguished. The former have been a bone of contention between secular and religious leaders for years. Each threatens the other, having no conception of the true nature and purpose of the recipe or the formula that was given long ago, which having been disregarded, has heralded in the universal decline of all Life.

Atomic energies are primordial – even more ancient than life on Earth. Without them there would be no quickening, animating energies. They come into being when solid, liquid and gaseous masses are accelerated in accordance with Einstein's equation:

$$E = mc^2$$

Strangely enough, it has never been stated that the acceleration inherent in this equation could equally be applied to functions that are in-rolling, formative and concentrative, as it has to those that are out-rolling, destructive and decentrative. If it is borne in mind that these two forms of motion are associated with two different types of atomic energy, then we begin to approach, or we are directly confronted by, the profound and veiled knowledge of Sais, whose disclosure according to ancient legend, was punished by death[76]. However, prophecies must be interpreted relatively. The death-ray can just as easily hit those who had

---

[75] According to Collins English Dictionary a bacteriophage is "a virus that is parasitic in a bacterium and multiplies within its host, which is destroyed when the new viruses are released." Here the meaning of *bacteriophagous* has little to do with bacteria as such, but tries to express the notion of highly subtle, autonomous entities within entities endowed with the potential to impart life- or death-decisive influences. – Ed.

[76] To give a greater understanding of events at Sais, that ancient Egyptian city in the Nile delta, the Secret Doctrine by H P Blavatsky relates: "...And how little Herodotus *could* tell is confessed by himself when speaking of a mysterious tomb of an Initiate at Sais, in the sacred precinct of Minerva. There, he says 'behind the chapel is the tomb of One, whose name I consider it impious to divulge... In the enclosure there are large obelisks and there is a lake near, surrounded with a stone wall in a circle. In this lake they perform, by night, that person's adventures, which they call mysteries. It is well to know that no secret was so well-preserved and so sacred with the ancients as that of their cycles and computations. From the Egyptians down to the Jews it was held as the highest sin to divulge anything pertaining to the correct measure of time. It was for divulging the secrets of the Gods, that Tantalus was plunged into the infernal regions; the keepers of the sacred Sybilline Books were threatened with the death penalty for revealing a word of them." *The Secret Doctrine*, Vol II, p.396. – Ed.

cause to conceal truth. Atomic energies should also be seen as something relative. In consequence, all the nuclear physics and the methodology concerning work and motion presently taught in all technical colleges and universities will no longer be valid, and indeed should even be prohibited by law.

For every action there is a reaction. Therefore a rapid succession of activating impulses or chain reactions arise, whose characteristic effect solely depends on whether a given mass-motion is initiated by *pressure* or *suction*.

It is a known fact that all motion is conditioned by different forms of kinetic energy (pressure and suction). Hitherto unknown, however, is that a motion triggered by pressure gives rise to *reactive*, decomposive radiation. Its effect is to increase thermal activity, which in all machines is commonly referred to as *resistance*. In motion of this kind, the resistance to motion increases by the square of the centrifugal acceleration of the rotational velocity. Such motion is therefore not only irrational, but it also impedes growth and development. Moreover all centrifugal systems of motion of masses, where pressure and heat increase, results in the molecular dissociation of the masses moved in this unnatural manner. This explains why biological, ecological and economic decline is occurring in those countries where academic (technical) methods of mass-motion prevail.

On the other hand: if movement is triggered by traction or suction, then an increase in power and efficiency results. This is due to the cooling, reactive forces evolving during this dynamic process, which manifest themselves as *levitation*. Levitation is a forward and upward surge which is intensified by reactions and founded on processes of molecular revalorisation. The resultant product is *qualigen*, which is the basis of all natural reproduction and higher evolvement, and the increase and ennoblement of all growth. There can be no doubt that the type of motion inaugurated, and the *reactive forces* (dynamic, atomic functions) associated with it, determine whether economic growth or decline, progress or regression, happiness or misery, war or peace, are to follow.

According to ancient knowledge, the world can be turned upside down and the natural course of evolution totally perverted if a retroactive motive impulse is given. However, if the correct, opposite impulse is imparted and the resultant motion is allowed to proceed unchecked, then equally devastating catastrophes will occur. Referred to in the Bible as 'The Flood', the Atlantean cataclysm was probably triggered off in just such a way through the over-stimulation of levitational forces (atomic, upsuctional forces), which ripped whole sections of the Earth skywards and atomised them.

The action of atomic energies that unleash the sacred fire in aqueous bodily fluids are substantially different. The ancient Jews called it 'Greek Fire' with which they set fire to sacrificial animals by sprinkling them with this energy concentrate before sunrise. According to the age-old scriptures, as the Sun rose, an almighty fire flared up that consumed the sacrificial beasts. The primitive people then sank to their knees and stammered out prayers of thanksgiving, for the fire was a sign that Jehovah had accepted their sacrifice. With these supposedly magic arts the high priests of both Indo-Germanic and Mosaic cultures ruled over the masses in order to force them to work and pay taxes. Just how the tax-raising financial power-structures came into being will be apparent to all who can read between the lines of what has so far been stated and what follows.

In bygone eras secular and religious knowledge resided in the same hands, because in those days savants were thoroughly versed in the art of controlling *both* kinds of atomic energy. They were thus able to regulate the supply of food and raw materials with the use of reproductive and upwardly evoluting forces. However, they did not let them get out of hand. These erudite men were likewise aware of the necessity for purifying forces. They knew that without a natural, healthy stockpile of bio-ecological residues and their naturalesque interment, none of the vital forces could be brought forth that enable the constant increase and qualitative improvement of posterity.

Without intuition (that extraordinary connection with Nature) this sublime art of controlling atomic forces is impossible. For this reason the final effect of bio-ecological disturbances could not be foreseen and we are now experiencing the consequences. In other words, this is simply the foregone bio-ecological conclusion of an intentional or unintentional suppression of life-giving forms of atomic energy. Because no one knew of the essential nature and the genesis of low-grade, pressure-creating, atomic forces i.e. decomposive energies, grave errors in motion could not be appreciated for what they were, and therefore continued to be made by the whole of science. Like a team of horses, science was used to bring about the intended developmental standstill in order to facilitate the regulation of supply and demand, and thereby the value of money, as desired.

This import and export trade in basic commodities, coupled with occasional shortages in other necessities of life, could only be manipulated as long as there were sufficient surpluses of goods in high demand. Therefore the need arose for a variable monetary value and rate of exchange. Through colonisation the machinery that unleashes decomposive energies gradually spread to less-developed countries: to areas inhabited by people whose way of life was still natural and who, up to then, had dwelt in relative freedom and independence. They took great satisfaction, even if unconsciously, in the progressive vitality of their various kinds of produce. Like wild game on the free and open prairie, they knew neither hardship nor stimulants of any kind.

In order to open up new markets in these regions too, what could have been simpler than to recommend these supposedly labour-saving machines to these 'primitive' peoples? From then on, the faster the heat and pressure-sensitive media of earth, water and air were forced to move unnaturally, the more dangerous the cell and tissue-rupturing, de-animating forces became. These then triggered off all-consuming fire and cold inflammation.

Nobody seems to have known or had any idea that if everything is relative, then motion must also be bipolar in nature. In view of this, Einstein's Law of Equivalence between Mass and Energy is an error of the gravest consequence, because it entirely depends on the type of motion, whether life-affirming or life-negating atomic energies, surplus growth or mass privation ultimately prevail. The energies that have been produced so far are the same as those that physicists generate in a split second in nuclear bombs, except they work far more slowly. The delayed effect of these decomposive energies only manifests itself much later, and as a result will perhaps no longer be explicable to future generations. It would appear that both Eastern and Western leaders are still ignorant of these two different forms of atomic energy. They are therefore forced to struggle for supremacy, threatening to make use of the weapons of annihilation that science has made available to them.

The spread of *centrifugal* methods of exploiting atomic energies will have even further unforeseen consequences. Even the lives of the high and mighty will be in doubt, when one day hydrogen bombs begin to rain down from radar-controlled aircraft and rockets. The point of this undoubtedly bad joke is that up to now nobody had any idea who the real joker is. He kept well out of the limelight and simply decreed laws that are reminiscent of the ancient oracular arts, because they can be interpreted in two different ways. This joker even went further and while promulgating a set of immutable and universally valid laws, at the same time declared that the exact opposite – relativity – also conformed with natural law. What is more, he did not even conceal the fact that all the constellations in the cosmos are moved planetarily. In doing so, however, he also made a blunder which, on closer examination, could bring about the collapse of the whole edifice of academic science.

As a spherical shape, were the Earth *de facto* to move centrifugally, then there would be no biomagnetic, longitudinal axis and no attractive or gravitational force. Without exception, these would have been fallacies for the simple reason that, due to the uncontrollable interactions arising from constantly-fluctuating temperatures, no state of equilibrium is possible. These interactions can only be controlled by lowering the temperature through the *centripetal* movement of mass. A state of equilibrium can then be maintained, and although not wholly constant, it nevertheless produces a steady succession of chain-reactions which serve the build-up of higher quality atomic energies.

The fact that centrifugal systems of motion of masses provoke a feverish, diseased condition in all living organisms is today beyond dispute. Modern technology and its perverted, unnatural processes therefore represent the deliberate application of a deception designed to make the working population unconditionally dependent on financial power-brokers. Moreover it has produced such an estrangement from Nature that the point has now been reached where students in technical colleges and universities are taught laws that neither exist nor apply in Nature.

Professor Hahn's achievement of splitting the atom follows the same sequence of events that occurs in every process of physical disintegration. A negative concentration of matter is bombarded with unfiltered positive rays, during which negative fructigens release more energy than the radiation or electron-guns are able to use. According to the type of radiation, these can penetrate all resistance and pierce right through to the cell-nucleus (atomic nucleus) and split it apart, triggering an unstoppable process of regressive development.

The Implosive Process of Breathing

All life breathes. It can also be said that *all that breathes is alive*. Breathing is controlled not physically, but metaphysically. When breathing ceases it means that the functional force, here called *life-force*, has been extinguished – upon whose continuing existence all respiration depends. Life-force is synonymous with the higher-grade, animating atomic energy that serves life.

All this and the following would be easier to understand if the process of breathing were to be described as the pulsebeat of a life-energising motor. If preceded by an explosion, a powerful vacuum develops and fresh fuel rushes into the void thus created. Exactly the same thing happens with *implosion*, which arises when a reduced fructigen originating from the Earth (negatively-charged) binds or emulsifies the seminal substances of the atmosphere. The product of this emulsion is an energy which animates all things.

Without an implosive process all breathing is impossible, and an air-pump would be required instead. Every living thing has its own individual body-temperature, which in humans lies at about +36.6°C (97.9°F). Any change foreshadows the onset of disease. At +36.6°C the body is able to emulsify in-drawn atmospheric oxygen and build up its vitalising forces. The significance of the Indian art of breathing (*prana-yoga*) therefore becomes quite apparent. This process of inner evolution or internalisation (implosion) can be produced mechanically and signals the nemesis of the greatest self-deception and betrayal of humanity, the technology that exploits retroactive reactions.

Processes of respiration that were previously unknown will now be briefly described. Through the intake of food and the consumption of water, predominantly negatively-charged substances stemming from the Earth reach the digestive system. Here they are purified and exalted, and enter the hermetically sealed blood and sap circulation systems by way of diffusion. Under no circumstances do blood or sap ever move centrifugally in the body, as any observation of the blood and sap streams will reveal. This is further enhanced by the presence of trace-elements in the walls of all high-grade vessels which, as catalysts, play a decisive role in the emulsion of the transported media.

It is therefore quite apparent that the relative proportions and activity of bipolar pressure-inducing substances determine whether a drop or a feverish rise in temperature ensues; the latter leading to sickening and degenerative after-effects. If the characteristic temperature level suited to the continuing existence of the respective organism alters, then the vacuum, the prerequisite for breathing, no longer functions. Here lies the great yawning gap in contemporary physics. Nuclear physicists are thoroughly conversant with the hot and cold conflagration leading to inflammation, but they are unaware of the heatless, formative processes for producing progressive, reproductive and upwardly-evoluting (multiplying and ennobling) atomic energies. Everything breathes, even earth, air and water. All will inevitably asphyxiate if reduced fructigenic ethericities (food) are unable to bind (emulsify) diffuse seminal ethericities, so creating the vital biological vacuum which produces hunger in the digestive tract and a demand for air in the respiratory system (lungs, gills, *etc.*).

This is all conditional on the presence of the necessary basic elements. Had humanity not provoked a fever-stricken condition in them, these would still be available in sufficient quantity and optimal quality. Today we are faced with a situation where those in charge of forestry, water, land and energy resources continue to proliferate these destructive forces, and with them the spread of cancer. All systems for moving or using water with pressure-turbines and pumps, Pelton wheels, water-supply pipes, river engineering profiles and their respective alloyed substances are unnatural and therefore wrongly designed. By moving water centrifugally, abnormally high reactive pressures, as well as infra- and ultra-red radiation, are produced. These are invisible and undetectable and contrary to Nature. All solid, liquid and gaseous basic elements should therefore be moved in the same way that Mother-Earth gyrates about and within herself along her ideal axis.

Higher-grade concentrates of atomic energy radiate their surplus energies mainly in a vertical direction. Since negatively-overcharged, bio-magnetic resurrective (ascending) forces are active here, many other things are also drawn up in their wake. This is the case with high-quality fish, for example, which are able to regulate these levitational forces with their gills in such a way that they can stand motionless in torrential mountain streams or suddenly dart upstream. In waterfalls they can even float up vertically, if the falling water is able to in-wind naturally. This also applies to us, for we are not aware of the weight of our bodies if they are healthy and are under the influence of the right temperature-gradient. Only in old age, when levitational forces begin to diminish, do we seek warmth and have need of a stick.

Divine Service attains its fullest meaning when Nature's formative forces are strengthened by supportive activity. Humanity will fulfil its ordained purpose when each individual makes a small part of this Earth fruitful and creates the right conditions for personal and general evolution. The key to healthy and prolific growth is to increase the range and quantity of formative forces. People who are healthy, contented and free are peaceable, and peace itself will come about of its own accord, because such people will refuse to take up arms for the sake of the plutocracy.

Secret of the Atom – (Conclusion)

In the preceding discourse 'the age-old secret of atomic energy' was explained in the simplest terms. It was pointed out that, whether reproductive and upwardly-evolving atomic energies evolve from bipolar elements, or whether degenerative carcinogenic energies are produced instead, simply depends on the type of mass-motion of the media of earth, water and air. It was also mentioned that in technical colleges and universities, laws, theories and principles are taught which are founded on the centrifugal, Einsteinian law of energy-generation.

On the basis of knowledge acquired in this way, forestry altered the natural methods of planting, conservation and felling; at the same time transforming shade-giving and shade-demanding mixed forests into mono-cultures. In agriculture, farm machinery was introduced which was not only improperly designed, but unsuitable alloys were also used. Methods of fertilisation were also introduced which destroyed life in the soil. In science too, mistake was heaped on mistake. Rivers were hydraulically (centrifugally) regulated and mountain streams made to cascade in straight channels, thus eliminating any possibility for in-winding motion. Waste-water was discharged into these regulated waterways without first being decontaminated. Even in the power industry they are only aware of the explosive and expansive motive principle. Due to the rapid increase in chain reactions triggered by the action of acceleration-induced pressure, the more developmentally dangerous its repulsive forces will become.

All of which justifies the following question: *Why has all this happened?* The fact cannot be ruled out that humanity was led astray for speculative reasons. Despite the enormous damage that already leaps to our very eyes, this deceitful behaviour will not be discontinued without a struggle.

Life-Force and Animating Energies
From *Implosion* Magazine, No.71.

"Atomic energy is freed at the intersection of two temperature-gradients. Whether it has a formative or destructive nature is determined by the form of motion and the alloys used in the manufacture of any motion-producing apparatus. The character of each individual is determined by its particular inner climate. Every life-form has its prescribed anomalous condition of health, which enables the orderly reproduction of the species in question. For this reason the world of parasites increases with fever."

V. S. – (*Implosion* Magazine, No.71, p.12)

This discovery is founded on the knowledge that, from a bio-ecological standpoint, there are two different types of mass-motion and hence two different forms of atomic energy. The first is of an *animating* nature, and is therefore not only regenerative and evolutionally up-lifting, but also multiplying and ennobling. The second form of energy is *de-animating*, destructive and hence regressive. Both forms of motion operate along a common developmental axis and do so simultaneously. The form of atomic energy that ultimately predominates depends on the type of impulse this rhythmical, dynamic process receives. These two types of motion encompass the whole panoply and spectrum of life.

This newly-discovered art of regulation was known to the high priests of ancient cultures and practised by them. They understood how to control the eternal metamorphic flow (*panta rhei*) in such a way that supply of food and raw materials kept pace with growing population. It was to this skilful control that the then peaceful progress and high level of culture are to be ascribed. It was all lost thousands of years ago, however, because it was restricted to a small circle for power-political reasons, and therefore withheld from the public at large. As a result its vitalising effect constantly declined. This ancient knowledge and art of control was rediscovered through observations of the stationary trout over many years, which floats motionlessly amidst torrential flows.

At this juncture, an explanation will be attempted. *Regenerative atomic energies* come into being when the media (carriers of bipolar basic elements) of earth, air and water are made to move *planetarily*. In this case the various sedimentary substances inherent in air and water bind the basic element, chemically termed oxygen (solidified solar energy), which carries an opposite charge. Here the principal requirement is the creation of an hermetic seal which excludes all light, heat and air, and the movement of the above media in an *in-winding* way. In addition, a higher excitation takes place by means of special catalysts. During the process of purification/exaltation, destructive forces of evolution are also released. However, these should only be released in sufficient measure to remove any waste products. As faecal matter and such like, these must be expelled if unwelcome deposits (crystallisations) are to be avoided, or if the decay of residues themselves, while still in the respective body, is to be prevented.

This marvellous *reduction-exaltation process* has been disrupted by contemporary technocrats, because they were only aware of centrifugal methods of mass-movement whose nature is degrading and destructive. For this reason, and in conformity with natural inner law, the constant acceleration of a regressive form of development is inevitable in all branches of industry, forestry, agriculture, water and energy resources management. Those principally to blame for this are the lecturers in technical colleges and universities, because they show their students how to achieve apparently lasting results by activating Nature's dissociative energies. Having thus been wrongly programmed, these aspiring academics then develop engines and machines, locomotive or otherwise, which serve destruction. These lecturers, who have ignorantly inaugurated a social decline encompassing the environment and all fields of human endeavour, are therefore to be described as the true *brake-appliers of evolution*, or as the long sought-after causal agents of cancer.

Indeed, if one considers that through the use of unnatural systems of mass-motion, millions of their unsuspecting fellow human beings are being denied their most fundamental right-to-life, ultimately to perish with unspeakable suffering (even rotting alive), then no word is harsh enough publicly to castigate this fraudulent motion.

Life-force can only be maintained through the input of supplementary *animating energies*. These become aroused and activated in transformations that take place on in-rolling spiral paths. During this process a cooling occurs, which acts as a brake on the destructive effects of oxidation and heat, because it neutralises them. An anomaly state (+4°C for water) is also created which varies from organism to organism and presides over its existence or non-existence. The high priests of earlier cultures knew how to manipulate this 'climate' so ably that the anomaly state, the condition of temperaturelessness and feverlessness, was maintained in all forms of life or growth. The variability of temperature stamps every manifestation of life with its individual hallmark.

Whenever anyone falls ill, it is an infallible sign that their natural state of indifference or metabolic equilibrium can no longer be maintained. The departure from this normal condition is caused by under-cooling or over-heating, by genetically-impaired food, feverish water or polluted air. In a word, over-acidic energy-concentrates have been ingested. If a still, stationary body of water, for example, is too heavily bombarded by unshielded (unbraked) solar rays, or if flowing water is subjected to pressure as it moves, then it becomes over-acidic and develops an unhealthy potency which the organism cannot tolerate. The organism becomes sick, begins to rot and slowly dies as regression takes its course. In the same way that to date no one has managed to produce artificial seawater in which sea fish can survive, nobody else (apart from me) has yet succeeded in producing the nutritive liquid, correctly-constituted physiologically, otherwise known as springwater.

Nor has anyone so far noticed that mountain springwater can only come into being with the aid of planetary motion. Here various other contributing factors are also involved which create the particular environment necessary for the oxygen component to be bound or emulsified by activated sedimentary material. Emulsion signifies that the ethericities (or emanations) of raw materials with opposite potential are so intimately intermixed that a third entity is born; as happens in any other procreative event. In practical terms, a child is born which is either male or female, according to the way in which fertilisation occurs. It can also be a hermaphrodite, which through later influences of various kinds can metamorphose into a predominantly male or female life-form.

Mountain springwater comes into being when, under diffuse influences of light and heat and the presence of catalytic secondary radiation, geospheric and atmospheric forms of radiation are thoroughly intermixed. In this way the child, the physically first-born and blood of the Earth, water, is *ur*-procreated. This physically first-born decays and dies if the spring becomes over-illuminated and overheated (when exposed to direct radiation and sunlight) in other words, over-acidified.

If the blood of the Earth is moved in a predominantly centrifugal manner – in an unwinding movement from the central axis towards the periphery – then hitherto-unknown ultrared radiation evolve which trigger off additional heating effects and make oxygen both aggressive and destructive. The final effect of this is the creation of a devitalising energy-form which radiates in all directions, penetrates all resistances and invades the negatively-charged cell-nucleus. Having been warmed in this way the formerly healthy living cell ruptures and turns into an epicentre of decay, infecting all forms of growth and life in its vicinity, until they too become genetically diseased (impregnated with cancer).

An apparently harmless mechanical, physical or psychic excess pressure is therefore able to trigger off localised processes of decay in the form of tumorous growths. We can observe these phenomena clearly in over-illuminated and overheated species of shade-demanding timber. The enlargement of the annual rings, the increasingly spongy structure of the wood ('light-induced growth') will then be praised as a 'major achievement of forestry'. This is how the qualigen in monocultures is forcibly degenerated, and how young trees are infested with cancer. Having thereby been deprived of its reproductive and upwardly-evolving force, the following generation produces infertile seeds and makes the groundwater cancerous.

If we wish to transform seawater or over-acidic fresh water into genetically-healthy drinking water, then this most valuable national asset must be conducted in such a way as to enable water itself to build up new energies again. This relatively simple regenerative process produces not only genetically-sound drinking and domestic water, but also high-grade healing water that inhibits cancer. In fresh water thus produced even sea fish are able to exist.

If we wish to safeguard humanity and above all our children from a horrible future, today's technology must be replaced by an environmentally-friendly bio-ecotechnology, a sustainable ec^{2}otechnology.[77] The principle has already been established for which pioneering patent applications have been made world-wide, and which in part have already been granted.

It is now time to begin, for with the continuing decline in the quality of all living things, the mental abilities of each individual will also disappear and humanity will no longer have the capacity to reverse the process.

---

[77] 'Ec^{2}otechnology': Reflects Viktor Schauberger's axiom: 'C^2: *Comprehend and Copy Nature*'. – Ed.

The Mechanical Generation of Life-Force
Archaeus – Vis Vitalis – Ch'i
From *Implosion* Magazine, No.57 – Viktor Schauberger, Linz, April 1955.

In a scientific journal, the German Professor Otto Warburg asserted that he had solved the problem of cancerous growths. According to him, the cancer cell lacked oxygen. A healthy cell is provided with oxygen via the blood. It uses this oxygen to process nutrients. In this way it is supplied with life-energy – in other words, with animating energy.

Professor Warburg then made the surprising claim that the cancer cell uses little oxygen. Instead, it ferments nutritive substances into lactic acid. It thereby obtains a great deal of energy and as a result grows faster and divides more rapidly than do other cells. No further progress has been made in the treatment and cure of cancer, because no one realised that in accordance with the Law of Bipolarity, there are two forms of energy, the *decomposive* and the *formative*. For this reason the following questions remain unanswered: *why does the cell-body suddenly use less oxygen?*, and *why, with the lack of it, does the cell begin to ferment nutrients, thereby losing some of its life-force and turning into a cancer cell?* Even Nobel prize winner Professor Gerhard Domagk (1895-1946) failed to answer these questions. As a result it was concluded that there is no causal agent of cancer.

Why and under what circumstances does the cell receive too little oxygen? No contemporary science lecturer can answer this crucial question. Because of this and despite the millions invested in cancer research, cancer has remained incurable and continues to proliferate and attack it victims.

Which form of energy leads to life and which leads to death? The cancer spectre will be laid to rest once the timely provision of sufficient oxygen to the cell has been successfully achieved. The metabolic process will then be altered in favour of the build-up of life itself. This force will then work equally as incessantly as does the destructive force in the opposite case.

This concerns the existence or non-existence of all humanity. Cancer causes living decay. Unspeakable suffering is concealed behind the walls of clinics and hospitals. A cure for cancer is thus of major concern to all humanity, and above and beyond this, the concern of every single human being, rich or poor, minister or street-sweeper alike.

The curing and prevention of cancer should under no circumstances be left to science alone. Science has failed right up to the present day. Despite its privileged position, it has not only permitted the spread of cancer in all levels of society, but by using the force of destruction in the field of physics it has actually *provoked* the development of cancer. On these grounds science is no longer entitled to any further monopoly. The same chances and the same challenges must be given to outsiders, and the value of any claims should be judged by results. The final decision must ultimately rest with the citizen, who has to meet the costs through taxation, and therefore should also have a say in any application of treatments.

Why can we, or why must we breathe? Professor Ernst Ferdinand Sauerbruch answered this vital question in 1908. The term 'breathing' is to be understood as an attraction or an insuction of diffuse oxygen. This is only possible if a *vacuum* or a n*egative pressure* exists in the thorax between the pleura and the surface of the lungs. If this negative pressure is lost, due to the effect of excess atmospheric pressure or inflammation, then breathing becomes impossible, and the afflicted life-form inevitably suffocates.

Before explaining how this negative pressure evolves, which rules over life or death, we must examine *oxygen*. Everything that appears in Nature and is perceptible to our eyes and senses, is the waste-product of subtle, exalted energies – left behind, foot-sore and weary on evolution's upward path, it manifests itself as physical matter. All growth should be seen in this light, which as far as present concepts are concerned only occurs through the influence of heat. To put it more succinctly, oxygen is the *waste-matter or fallout of solar energy in gaseous form*. It is waste-material similar to the one we commonly call *nitrogen*. Together they form a mixture of gases – the atmosphere – in which, despite their unequal proportions and weights, a labile state of equilibrium prevails. This is controlled by constant fluctuations in temperature in such a way that a periodic drop in pressure and heat occurs and a low or negative pressure is created. The biological consequence of this constant alternation of potentials is a continuous equilibrating motion, whose causes are *indiscernible*. What is perceptible, however, is the way these causes manifest themselves through up- and down-draughts, whose roles are therefore already of a subordinate nature.

It is known that physically-detectable motion can be initiated through differences in potential. Hitherto unknown, however, was that motion can also be used to create differences in potential. The decisive factor in this regard is that the resultant equilibrating motion can not only be controlled, but also intensified at will. It can therefore be regulated mechanically in such a way that an upward suction-force evolves which draws everything in its wake, thus negating gravity. One day we will probably be able to explain the cataclysm of Atlantis in this way, of which we have a foretaste in cyclones. These originate through potential differences in the strata of the atmosphere and in the process of in-winding, intensifying reactive forces develop energies, whose combined power is equal to 150 hydrogen bombs.

An invincible force of gravity therefore does not exist, and solid, liquid and gaseous masses can actually become almost weightless if they are simply moved in a naturalesque manner, i.e. *planetarily*. This motion is in no way circular, but must take place in cycloid-spiral space-curves. In this instance a mysterious and *precisely measurable* negative pressure comes into being, which attracts and draws in so-called atmospheric oxygen with inconceivable force, processes it into nutritive material, and in this way, what Professor Warburg called *life-energy* evolves.

This life-energy immediately subsides if any medium (air, water, earth) is accelerated axially→radially by means of a drive-shaft. This unnatural form of motion can be described as life-negating or life-energy-destroying. With it all parasites are given *carte blanche*, for in Creation their function is to demolish all that is no longer evolutionarily viable.

With an increase in heat and pressure, the vacuum or insuctional force is lost. This manifests itself first in the more highly-evolved genital organs and in the brain as tumours. Whatever can no longer breathe, suffocates, and if its immediate environment is abnormally warm it begins to decay. Contemporary science is oblivious to all this, because it knows nothing of the processes and equipment required to move any given medium centripetally or along involuting spiral paths. The insuctional force thus created not only draws in higher-grade elements in trace form, but also generates energies that serve the vital functions. This emulsion also requires the co-action of catalysts. Without these binding energetic influences, no union between the counter-polar ethericities of the basic elements is possible.

Is there a machine with which life-energy can be created or produced? According to press reports, a university in California has been making progress with research into photosynthesis. Through photosynthesis, however, original (form-originating) potential differences between the positively-overcharged atmosphere and the negatively-surcharged geosphere can be created artificially, thus obviating the use of coal and oil for producing mechanical power. In order to achieve this, oxygen has to be bound by nitrogen by a cold process, which is to the benefit of the latter. This would result in the artificial production of natural protein concentrates, which are the principal building-blocks of Nature.

Just where would all this lead? It would enable the manufacture of *breathing* machines which rhythmically impregnate and artificially respirate all formative substances created in the process, and which generate primordial life-energy mechanically. It would also ensure humanity's freedom of movement on Earth, on and under water and in the air, for all time.

These elementary forces manifest themselves as catastrophes of all kinds, as thunderstorms and deluges, and in the tropics as cyclones and typhoons.

The task of up-and-coming scientists is to harness these forces for the service of humanity. The example most suited to explaining them is that of the stationary trout in a torrential mountain stream. Without practical examples and suitable illustrative material it is difficult to describe the vast differences between *implosion* and *explosion*. Were it otherwise, then the sense and purpose of breathing could also be explained biologically, which makes this *negative pressure* possible, provided the intermixture of the products of diffusion proceeds correctly.

In principle Professor Warburg is right when he states that it is the lack of oxygen that triggers off reversed metabolic interaction, whose products of emulsion rupture the cells. However, even this scholar, educated according to an old school of thought, overlooks the fact that children go blind and adults succumb to an incurable lung-inflammation, if nitrogen-less oxygen is introduced into the lungs under pressure. It makes a tremendous difference whether this gaseous waste-product of solar energy is *drawn* into the lungs or whether it is *pressed* into them. This could be compared, for example, to the difference in tone-quality between a suction- or pressure-driven harmonium. The tone of the latter is harsh, whereas the tone of the former is ethereal (of the spheres).

Today's technology has subverted the eternal river of transmutation (*panta rhei*) into a river of death, whose products of emulsion are forms of energy akin to those produced mechanically by today's nuclear scientists and engineers. In Nature everything is reversible. If a metabolic interaction is triggered off *implosively*, it produces the highest-grade upsuctional forces (levitation), which alleviate all intellectual torpor and negate all physical weight – as they also do in the case of the stationary trout. The age-old Aryan testament, the *Tabula Smaragdina* urges: "*Mix the stuff of the Heavens with the stuff of the Earth naturalesquely, then throughout life you will have independence and happiness.*"

This can all be achieved with the implosion-machine, which unwinds, inwinds and converts excess atmospheric pressure into a concentrative negative pressure. With these naturalesquely engineered differences in potential it is possible not only to produce driving forces for all growth, but also to generate them in properly constructed, alloyed and insulated machines almost free of cost. With this discovery humanity now stands before a choice: to take leave of a purely physical developmental paradigm, which threatens life with increasingly destructive forces; or, unconditionally to fall victim to the cancerous scourge of the 20th century.

A simple experiment carried out correctly, proved that it was possible through the construction on an implosion-machine, to generate the same life-energies on a large scale, which come into being when we breathe. The product of these is etherealised food.

Is There Perpetual Motion?

From *Implosion* Magazine, No.26. Written by Viktor Schauberger at Leonstein, July 1945.

*"Every movement consists of two components. One component serves inwardness (intensification), **the** other **serves** outwardness (dispersion). These two pre-requisites alone regulate the eternal river of metamorphosis (panta rhei)."*

V.S. – (*Implosion* Magazine, No.57, p.5)

Merely posing the question, *is there perpetual motion?* exposes one to the risk of losing the respect of one's fellow men. People have become accustomed to thinking that such a form of motion is impossible. It is raised here in order to examine it in connection with a variety of experiences, which point to the fact that perpetual motion does indeed exist. Not in the sense, however, that science has hitherto considered it, which logically has led to rejection of this often-debated question.

Above all we are concerned here with the question, *why does Earth rotate about its own axis?* The choice of words is actually not quite correct. The term own should be replaced by the word *peculiar*, stressing the uniquely particular axis about which Earth rotates. Even the meaning of the word *rotate* must be qualified, for the simple reason that everything that turns in circles goes nowhere and makes no headway. There is no change of position, which in the case of the Earth's motion is impossible. It is precisely this *making headway* that is the purpose of this strangely-moving Earth. This process of movement could be far better described in terms of a *pumping action*, through which our hearts are also made to 'beat', so it is said. The movement of blood, however, is not the result of the so-called beating of the heart – it is the cause.

If these fundamentals, which hitherto have been taken for granted, are reversed or inverted, then the scorned expression *perpetual motion* takes on a substantially different meaning. My opening question is then no longer absurd, but actually becomes highly interesting. In this connection other questions come to the fore: *what actually is life* and *how does life come into being?* A whole complex of questions thus attains a proper perspective and its final clarification is attempted here.

When Galileo discovered that the Earth rotated, it sent shock-waves through the secular, religious and scientific worlds, which resulted in his enforced retraction of this assertion. It is known, however, that before he died, Galileo retracted his retraction, and since then it has become obvious to all that the world rotates about its own axis.

So far nobody has made the effort, however, to reflect upon *why* the world-sphere actually *turns*. For various reasons even the word *sphere* conveniently conceals the function of Earth's distinctive motion. In two respects the organisation of variously weighted masses within the Earth in no way permit the formation of a true sphere. Apart from absolute and specific dif-

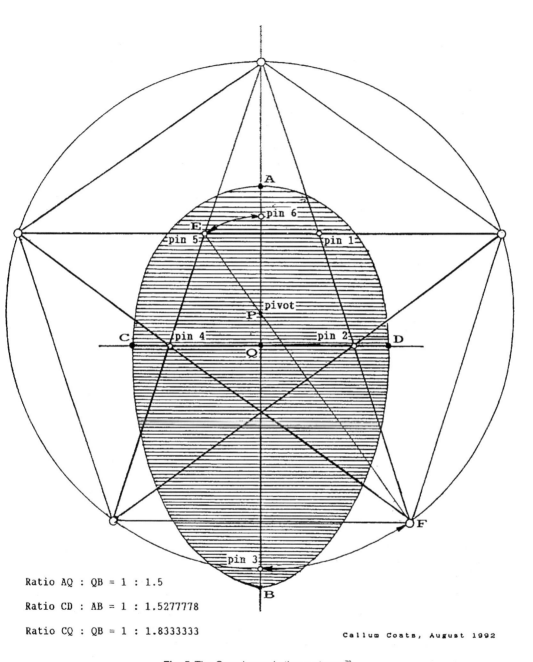

Ratio AQ : QB = 1 : 1.5

Ratio CD : AB = 1 : 1.5277778

Ratio CQ : QB = 1 : 1.8333333

Callum Coats, August 1992

Fig. 5: The Cosmic egg in the pentagon[78].

[78] Viktor Schauberger has not explained how this egg in the pentagon is derived. This diagram, which may not be correct, is reproduced from the editor's own book, *Living Energies*, to show how an egg can be drawn, based on the geometry in the pentagon. – Ed.

ferences in the weight of its internal substances, different directions also come into question, for some of the constituents react to centrifugence and others to centripetence. All of this has been proven unequivocally by experiment.

This fact alone demands an *eccentric* form of motion, which in turn is mirrored by the asymmetrical shape of the Earth. The form in question is the *egg-shape*, and more particularly, the 'extreme' egg-shape, which can only be derived from the five-sided symbol of the pentagon (fig. 5). It provides the basic structural framework to deal logically and self-evidently with all the peculiarities to be presented. Considering this peculiar type of motion, this peculiar moving body and above all its peculiar axis, we can ask the question, *why does Earth move in such a singularly distinctive manner?*

This whole affair is akin to one that led to the most serious errors in the field of scientific experiment. A given experiment is carried out a thousand times and always produces the same result. The reason for this irrefutable result is the repeated use of an *identical precondition* under which the experiment is conducted. If this is changed, then obviously a different result is obtained, which may have been neither sought nor desired. In this case there are absolutely no grounds to be able to speak of a regularity, but rather of a *fluctuating conformity* with natural law. This is understandable, because the extremely crucial influences of light and temperature, and indeed all other types of influence, are constantly changing. In Nature therefore, the preconditions for such experiments are totally absent, whereas in a laboratory in which specific preconditions can be artificially arranged, this or another experiment may be successful.

From this point of view, any such experiment is misleading, as is Archimedes' Law of Displacement. If water is moved naturalesquely (in the way that healthy water has to move, analogous to the form of Earth motion above) before Archimedes' specific-weight determining experiment is carried out, then its specific weight changes so markedly that the laws governing the definition of specific and absolute weight are thrown to the four winds. In this case ores, which would normally sink to the bottom at once, actually float in the central core-water of a river. Even more amazing is the behaviour of heavy stones in still water, which under particular conditions of temperature no longer obey the law of absolute gravity and rise to the surface. However, this only happens if the water has previously been moved according to this natural (peculiar) pattern of movement. All of which means, no more and no less, that the law of gravity exists only under certain preconditions.

If we continue further in this direction, always mindful of the question of preconditions, then through the particular form of motion described in general terms above, the *law of gravity* can be eliminated. A completely new law of levitation becomes operative. This throws the whole of science and technology overboard, because up to now these two fields of human endeavour only knew and applied the law of gravity.

Gravitation, however, is only a secondary effect of this principal force. Were there no levitation, then gravitation could never exist, because in the final analysis everything would remain flat on the ground. By using this levitational force, the energy to power machines will be virtually free and therefore of little worth by today's ethical standards. From an environmental point of view this makes good sense, because it serves the will of Nature and furthers evolution. In contrast, present systems of energy-generation using the forces of gravity and pressure are environmentally dangerous, and the more they are exploited the more hazardous the dangers.

Having said all this, however, we are still just at the beginning of the introductory description of this peculiar form of motion. We are also at the initial phase of the question, *Why does the Earth move itself?* In answering these questions we inevitably get closer to the resolution of the question, *Is there perpetual motion?* Certainly there is, but in a completely different sense to the one that has so far persuaded science to dismiss the matter *a priori*, because at the very outset nobody should concern themselves with such nonsensical things. All of this, despite the accepted fact that, in ecological consequence of the above 'peculiar' form of motion, the whole Earth *floats* in free space without an axis and hence without friction. This is due to its movement along a *cycloid-spiral space-curve*.

This *singularly unusual* axis of the Earth is therefore *no axis at all* in the accepted sense, but an infinitely small hollow space or, more accurately, a concatenation of an immense number of hollow spaces akin to pores. In these spaces not only are the products of levitation to be found, which come into being through this curious motion, but also gravitational products arising from the planetary counter-motion of the Sun. When viewed from the same vantage point these two forces rotate in mutually-reversed order within and without this remarkable evolutionary form (Earth). Even the words 'reversed order' are not quite correct, because the end-effect results in their moving in opposite directions (upwards or downwards, inwards or outwards) owing to their mutually-inverted direction of rotation.

Apart from these differences account must be taken of differences between *active and reactive forms of temperature*, of the different directions taken by the *actively-reactively functioning products of temperature*, and also, the various types and directions of light (which again produce various differences in potential). The outcome of the deliberations highlighted here will be of such diverse nature that we stand before an entirely new world. Contemporary ways of looking at things thus become totally unworthy of consideration. The ramifications are of such a staggering nature that we have to start again at the very beginning in order to arrive, in roundabout ways, at the point where life is once more worth living. It is in this sense that what follows should be considered.

However, this is not so simple because bipolarities have to be considered. On closer inspection we must also take reactions into account, so that we must speak of a *tri-polarity*. Then we can differentiate between the products of spacial organisation and those which in effect do not manifest themselves as pure bipolarities, although they are the cause of this regenerative and upwardly-evoluting form of motion. The end-products of this phenomenon are specifically-densified forms of the structures of raw and refined matter, which rhythmically reinforce and raise each other up and qualitatively enhance each other. New interactions take place, resulting in an eternal, unceasing motion. This leads to the realisation that there would be absolutely no life or movement were there no perpetual motion in Nature.

It is difficult to find words to describe this 'peculiar' evolutionary production line upon which the quadratures of this 'original' metabolic process arise, and the cubatures of the raw materials take form. These should be visualised as looping skeins of endlessness akin to the mathematical sign for infirnity – ∞.They rotate about and along the longitudinal axis in such a remarkable way that the centre of gravity of a through-flowing mass is so guided along the axis that the 'disappearing' encounters the 'returning'. They become so entangled with each other in the process that the disappearing entwines (binds) the returning, or *vice versa*. In the first case a process of *specific* densation and an autonomous upsurge or upward impulse takes place, which happens with all good mountain springwater. The opposite case results in a process of increasing *absolute* density in a gravitating fall and repulsive recoil.

It is upon these differences that the genders of various entities or formations depend. For this reason, whatever has been created in this way has to interact continually. Through the agency of a special impulse-and-repulse generating motion, the crude material cannot disappear, but by being moved *implosively* it is continually reborn in the production of refined matter (ethericities). Owing to its polarity it must also interact, thus initiating a new motive impulse. One could become quite demented if one reflects not only upon this eternal, dynamic process, but also upon the resulting products of motion! This would be the case if a certain limit were exceeded. The ultimate manifestations in ethereal and spiritual realms can no longer be contemplated because one has to keep one's feet on the ground.

In this 'peculiar' axis, which can be perceived with the naked eye as the tube-like structure that forms when water flows down a drain, the source of levitative matter is to be found. In this same axis the trout is able to overcome a 10m high waterfall without effort. He who understands how to cultivate this levitating force by mechanical means can create 'perpetual motion', because in their bipolar interaction the levitating and gravitating forces produce a repulsive, uplifting motion.

The same bipolarity also operates in the realm of the mind, wherein *intuition* is to be found in the upper regions and *logic* in the lower. Similar differences exist between the conscious and unconscious mind, which in most people has become so atrophied through one-sided education that the *gift of intuition* is now limited to a very few. Whatever we create using our free, over-developed, intellectual capacity is mostly upside down. Logical thought is only the initial phase or the prerequisite for biological action. For this reason it is also understandable why secular and religious leaders were so rudely shaken when Galileo first mooted his extremely ticklish question. At about the same time Leonardo da Vinci began his search for *il primo motore*, which eternally moves our Mother-Earth about her special axis. Nothing could have been simpler or more obvious than to copy this 'peculiar' motion of Nature's faithfully in *Repulsators* and *Repulsines*. The logical outcome of these unique experiments then produced the naturalesque preconditions, which enable the generation of energy for machines, as well as all the energies for growth of whatever kind, at virtually no cost.

In accordance with the maxim therefore: *"First think forwards (intuit), and only then think backwards (deliberate), for intelligence and intuition are also inverted"*.

The Motion of the Earth – The *Ur*-Cause of Radiation – The Resurrection of Life

An amalgam of two short papers by Viktor Schauberger – Leonstein, July 1945. – Schauberger archives

Galileo Galilei discovered the revolution of the Earth about its axis. This discovery provoked an immense furore in secular and religious science. Gallilei was forced to forswear, but shortly before his death, however, retracted his enforced retraction with the world-famous words:

"and all the same, it still rotates!"

Why this was so shocking has even today not actually been made clear, and more particularly, why Gallilei was forced to renounce, to recant. All manner of observations and investigations of many years standing led to the realisation, however, that the arousal of the secular and religious leaders was not without foundation.

The revolution of the Earth about its own axis is really an oscillation of an *egg-shaped*, internally variably-weighted, hollow form, above which arches the horizon. This motion of the Earth is the *ur*-cause of radiation or radiance, the emergence of that we call 'life'. Through an oscillation along two different axes, a concentration arises out of a bacteriophagic nebula. At the common focal point of these two axes, this concentration then produces a life-quickening expansion.[79]

Due to this centrifugal gyration, analogous to a spinning top, two mutually opposed dynamic components come in to being. (see fig. 6a)

1) A centripetal striving to centre itself along its longitudinal axis.

2) A figure-of-eight shaped horizontal motion (viewed in transverse cross-section – see fig. 6b), which due to the unequal weight distribution, either reduces its wobble with increasing rate of rotation, or increases it with a reducing number of rotations.

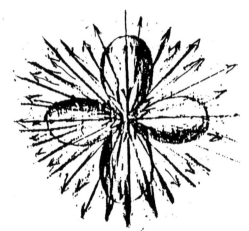

fig. 6a – Longitudinal View   fig. 6b – Transverse View

Fig. 6 The energetic interaction of centripetal and centrifugal forces. (see also fig. 7)

If the number of revolutions rises, then the self-generated oscillation arising from the horizontally looping motion contracts, resulting in the automatic reduction in the rate of rotation.

If the number of revolutions diminishes, then the figure-of-eight shaped outward oscillation above and below the horizontal axis increases in length at the point of largest radius. In this case, the previously declining distance from the ideal, *magnetic* axis increases, resulting in a renewed intensification of the rotational force, which in turn increases the rotational velocity.

[79] This phenomenon is akin to the energy-expansion points shown in figs. 4, 5, 6 & 13 in *The Water Wizard*, vol.1 of the Ecotechnology series. – Ed.

THE ETERNAL "CENTRIPETENCE"-"CENTRIFUGENCE" CYCLE	Longitudinal View Biomagnetic Axis	Transverse View Bioelectric Axis

At 4: The egg is at its minimum rotational velocity about the longitudinal biomagnetic axis, maximum radius and maximum centrifugal condition caused by the maximum centripetally-induced elongation of the 4 transverse figure-of-eight loops, which are at their maximum spin. The egg's 2 longitudinal figure-of-eight loops are at their minimum spin, minimum biomagnetic state, maximum radius and maximum centrifugal condition.

At 1: The 4 transverse figure-of-eight loops are at their maximum spin about their transverse bioelectic axes, minimum radius above and below these, maximum elongation, maximum spin and maximum centripetal condition.

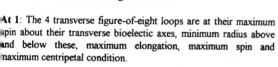

At 3: The egg's maximum centrifugal state having culminated, it begins to contract centripetally, to rotate faster, and its 2 longitudinal figure-of-eight loops to elongate along the biomagnetic longitudinal axis as they contract centripetally and spin more rapidly as the biomagnetic force increases.

At 2: Their centripetal state having now culminated, the 4 transverse figure-of-eight loops begin contract along their bioelectric axes, to expand centrifugally above and below these, and to spin more slowly as the bioelectric forces decreases.

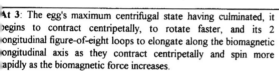

At 2: The egg is at its maximum rotational velocity about the longitudinal biomagnetic axis, minimum radius and minimum centrifugal condition caused by the maximum centrifugally-induced contraction of the 4 transverse figure-of-eight loops, which are at their minimum spin. The egg's 2 longitudinal figure-of-eight loops are at their maximum spin, maximum biomagnetic state, minimum radius and minimum centrifugal condition.

At 3: The 4 transverse figure-of-eight loops are at their minimum spin about their transverse bioelectic axes, maximum radius above and below these, minimum elongation, minimum spin and minimum centripetal condition.

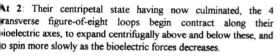

At 1: The egg's maximum centripetal state having culminated, it begins to expand centrifugally, to rotate more slowly, and its 2 longitudinal figure-of-eight loops to contract along the biomagnetic longitudinal axis as they expand centrifugally and spin more slowly as the biomagnetic force decreases.

At 4: Their centrifugal state having now culminated, the 4 transverse figure-of-eight loops begin elongate along their bioelectric axes, to contract centripetally above and below these, and to spin more rapidly as the bioelectric forces increases.

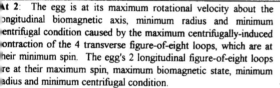

The cycle now completed, it begins again as in the top diagram. This interaction between biomagnetic, centripetal and bioelectric, centrifugal forces happens so quickly that it appears as a state of rest. Where the axes of these two opposing, but complimentary forces intersect, life-energy is emitted.

Interpretation by Callum Coats[80]

Fig. 7: The eternal centripetence-centrifugence cycle.

[80] Not easy to understand for either of us, I discussed these combined papers at some length with the late Dr Tilman Schauberger, Viktor Schauberger's grandson and expert on his grandfather's works, who gave them to me in late 1988. **Fig.7** is the result of these discussions and is submitted here as a possible aid to better understanding. – Ed.

In a graphical depiction, there are two figure-of-eight-like loops, each of which, rotating about its own axis, triggers off a bi-axial spiral oscillation. These looping movements are both mutually antagonistic and mutually complementary, and culminate in the creation of a 'labile' unstable state of equilibrium, a state of floating, whose inner fulcrum is the focal point mentioned above. At this point the concentrating and expanding forces intersect, which, acting in all directions, produce the 'labile' point of balance in which two symmetrically and proportionally reciprocal basic elementary states alternately consume or bind each other. This unleashes the life-giving motion, which oscillates radially in all directions and once more binds the products of expansion in the same focal point or point of concentration thus brought into being.

This process of motion, however, is only possible in an extreme egg-shape derived from the pentagram. This is isolated in such a way that from inside outwards and outside inwards, an interaction between potentiated bipolar expansive and concentrative substances is possible. Dispersed partly mechanically and partly by reactive temperature forms, these enter and exit through reciprocally disposed pore-openings equipped with bipolar diffusion coils. These generate the influences that underlie the inward and outward falls mentioned above.

If one takes into account the contrasting nature of the substances of the Earth and the Heavens, some of which react to centrifugence and others to centripetence, then there arises a constant mixing of both opposing formative basic elements. This produces a rhythmical, irregular combination of both oppositely orientated stocks of basic elements, and a regular disturbance of the bi-axial state of equilibrium, resulting in an infinite curve where in practical terms there is no longer any state of rest. This infinite motion is the *'cycloid-spiral space-curve motion'*, the secret of *'Eternal Life'* through *'Eternal Motion'*. And *vice versa*, eternal motion through eternal life.

In this concentrated form of representation, it is impossible to comprehend the expansive action of this specific ur-cause-concentration and for this reason one must go to very great lengths in order to describe the focal point of this unique ex-pansive concentrative motion.

The Secret of the Egg-Form
From *Implosion* Magazine, No.112, p.56, written in Vienna October 1940.

When a drill rotates at about 28,000 rpm, the drill-bit can be made to bend without breaking by displacing the material being drilled. When the high-speed drill-bit is bent, heat is generated at the point of flexure, the position of the smallest radius at x (see fig. 8a). Where the drill-bit is not under load, the faster the rate of rotation, the thinner the shaft can be. The thinner the shaft the less the driving force required, i.e. with the thinnest shaft only minimal boring pressure is possible. The slightest overloading causes the drill-bit to break.

Through this observation I have become aware of a hitherto unknown source of energy, which could fundamentally change current technology. Instead of transferring power directly, it is possible to exploit reactive forces obtained by indirect means.

If air or water is centrifuged[81] in a naturalesquely constructed egg-form, then only a portion the material conglomerates or reacts to the horizontal centrifugence. This is because these organisms (air and water) have been endowed with catalytically active character through the addition of bipolar minerals, and catalysatorically active character through the incorporation and arrangement of bipolar wall-surfaces. That is to say, only the carbones and hydrogen are conducted to the position of greatest radius. There, with rising pressure due to increasing rate of rotation, they are forced against the rotating inner wall of the wobbling egg-form (see fig. 8b).

The *'higher'* waste matter, so-called oxygen, separates out and accumulates about the longitudinal axis of the egg-form, the latter rotating about its own axis. The higher the rpm, the smaller the cross-sectional area of the oxygen core ranged around the longitudinal axis (shown at * in fig. 8c).

Now an intermediate phenomenon:

Through the increased rpm-induced rise in pressure, a potential or charge develops at the position of greatest horizontal radius. After a certain critical minimum pressure has been exceeded, this gives way to a depotentiation or discharge, which results in the relapse of the water previously raised through centrifugence (see fig. 8d). As a counter-effect, the release of an ampère-less energy-form can be observed, which transpierces the wall at the position of largest horizontal radius, and flows away in a wavi-linear manner (fig. 8d).

After a short period the de-energised and thus previously relapsed water rises once again to the point of maximum radius (see fig. 8e). Immediately after the water has subsided, the rate of rotation rises. When the relapsed water again rises at the location of maximum horizontal radius, then an exponentially increased pressure is exerted on the wall-surface. As a result the partially de-energised water can once more release energies, which again transpierce the walls at the point of maximum horizontal radius. Whereupon the newly de-energised water once again sinks. At this instant, the rate of rotation again increases, whereby the lateral pressure again rises exponentially, and so on.

---

[81] A handwritten note by Viktor Schauberger states that the effect occurs at 32,000rpm. – Kurt Lorek

The 'higher' waste matter, the oxygen, executes a rhythmically ordered interplay and every time the water sinks, the oxygen converges further towards the longitudinal axis, reducing the cross-section of the oxygen core in the process (see fig. 8f). In other words, every de-energising of the carbones and hydrogen on the horizontal axis (plane) results in the charging (potentiation) of the longitudinally-centred oxygen core. That is to say, with every lateral depotentiation, the oxygen core is extended longitudinally (vertically), its tip developing towards the top of the longitudinal axis.

The higher the rpm rises in consequence of the rhythmical depotentiaton of the carbones and hydrogen (at which time the ammeter indicating the amount of current falls), the more elongated the self-centring oxygen core becomes, due to the heightened rpm. This centrally disposed, increasingly slimmer and longer egg-form is the 5th egg-form (see figs. 8c & 8g), which is formed within a 2nd, 3rd and 4th egg, each of which encloses and surrounds a more highly cultured substance. Through processes of atomic fission, therefore, five zones are created, the point of the innermost of which also points towards the narrow end of the outermost egg-form. When the maximal velocity of the rotating outer egg-form is exceeded, ions stream out of the tip of the innermost egg. These are emitted in pulsations and are shot into space at extremely high velocities.

If these ions, which with concentric compactness rotate about their own axis, encounter a twisted conductor (see fig. 8h), then this pure kinetic energy is transformed into heat and consumed in the process. This heat-form, however, has a contracting force and sinks, in contrast to the familiar diffusing heat-form, which rises. The first heat-form, however, is important for the formation and ordering of growth although this is only incidental.

If these ion-discharges are conducted to the exterior via the points of suitable cones, we are presented with a kind of ion-cannon (see fig. 8i). With this the radius of action and the velocity of the emitted ions can be regulated at will by increasing the rotational velocity of the egg as it rotates about its own axis. When an emitted ion encounters a substance floating in space at high altitudes the substance will be heated up, de-energised or discharged in the process and disintegrated. With every ion-emission, therefore, the particles (micro-spaces) floating in space will be reduced in size. These substances will be bombarded until even the original micro-space becomes space-less or reverts to a pure energy-form.

Since this energy-form carries an opposite charge relative to the emitted ion, any following ion carrying a positive charge will be attracted to the negatively-charged products of disintegration, due to its opposite polarity. In this way the original force of emission will be intensified through elementary forces of attraction.

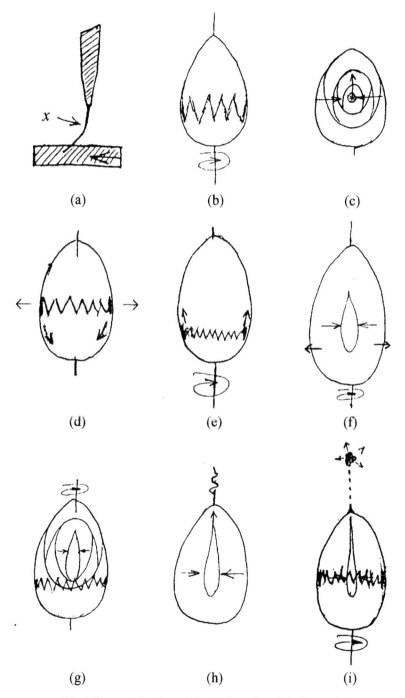

Fig. 8: The centrifugal – centripetal dynamics within the egg-form.

Further conclusions will not be addressed here, since these rightly belong to the processes of formation itself. These provide the impetus and basis for growth in the form of counter-ions, which come into being through the reversion of the highest energy-products. Growth must therefore be considered from a substantially different point of view than it is today.

Vertical winds, cyclones, typhoons, hurricanes, waterspouts and so on could be viewed as natural examples of this, which is why the Repulsator is an ideal pump. The Repulsator, for example, can suck in seawater, atomise and draw it upwards. The water's content of oxygen also does not react to centrifugence, but operating in the opposite sense, shapes the movement on the longitudinal axis, its end-product is actually the positive ion. For this reason the remaining water centrifuged out through the waviform is sweet (fresh) (see patent example).

That the formation of the Gulf Stream and the movement of sea and air currents can also be explained in this way, is self-evident. In this way too, and for the first time, the *ur*-formation of the Earth, the *ur*-creation of water and other zone-groups, the *ur*-creation of the bio-anode of the Moon and naturally the *ur*-formation of the bio-cathode of the Sun, becomes understandable. The secret lies hidden in the '*own*' axis, which comes into being through the atomic fission and centring of the oxygen core and the dynamic motion of a naturalesquely constructed egg-form.

Vienna, October 1940.

Organic Syntheses

Organic Syntheses

Processes and equipment for carrying out organic syntheses with the aid of droppable liquid or gaseous elements. From *Implosion* Magazine, No.22

Today's technically-minded people have already divorced themselves so thoroughly from Nature that they are simply incapable of understanding such terms as *organic syntheses* and *plasmolytic forces*.[82] For this reason the following elaborations are necessary as data for later practical demonstrations, otherwise nobody would be able to understand how to build the necessary devices and equipment. The same situation would then arise as has occurred with contemporary methods of generating electricity, already carried out on a large scale without anyone knowing what the concept 'electricity' really means. Because of this no-one is aware of the ultimate after-effects of this force, which in resistance creates *heat* and in this energetic form conduces *decomposition*. We have therefore set out on a journey whose end and final outcome we cannot know.

In contrast to degenerative, decomposive 'electricity', bioelectricity generates fresh, renewed life – growth. We are therefore concerned with extreme differences and with equipment and appliances which transform the organic remains of life into a *'biocurrent'*. It is a process akin to the way that plant-fruits, grapes, a tart apple, or a cancerously acidic perry-pear can be transformed into *spirit* (alcohol). This arises through the varying temperatures active in a process of fermentation. The reorganisation and higher synthesis of the residues of human, animal and vegetable life depend on causes

---

[82] 'Plasmolytic': Collins English Dictionary describes 'plasmolysis' as "the shrinkage of a protoplasm away from the cell walls that occurs as a result of excessive water loss, esp. in plants cells." The implication here in Viktor Schauberger's use of it, is that it refers to a form of motion that withdraws or elongates the moved medium (earth, air or water) from contact with the conducting wall-surfaces. This reduces all resistance-, friction- and heat-producing processes to a minimum. In addition it also withdraws the medium into itself, thus concentrating its molecules into a an increasingly smaller volume. This internalising process also extends to the withdrawal of the energies themselves from the walls, as it were, of the confining molecules, leading to their implosion, a reversion to an atomic state and a super-concentration of pure energy. The end-result of this is the 'biological vacuum' and the phenomenon that Viktor Schauberger referred to as the 'compression of dematerialisation'. Instead of being a void in the usual sense of the word, the biological vacuum is actually a superordinate state of extraordinarily high potency.- Ed.

opposite to those that trigger such high-grade decomposive processes. Expressed more accurately, they depend on metaphysical processes of further development. These are inexplicable in terms of present concepts and outlook.

Broadly speaking, what needs to be addressed here is a practical, implementable process of atomic transformation (not fission) with the aid of life-current or biocurrent-generating biomotors. With these growth can not only be increased at will, but a wide variety of machines and motive power can also be produced for virtually nothing. In this new world shortages of food will be a thing of the terrible and grisly past. Energy will be free and therefore of no commercially exploitable value.

Today this train of thought would appear utopian. All the more so, because it is exceedingly difficult to differentiate contemporary concepts from those required here, for which there are no technical terms. Therefore modern (and frequently inappropriate) terminology will have to be used. By placing them between quotation marks a different sense or meaning is intended. This I can do nothing about. Later on, other and fresh words will have to be coined to describe these new concepts.

The False World View

In every larger settlement there is a church and next to it a school. In the school the teacher works and prepares the rising generation for their professional and adult life. This rising generation is evolutionally older – they enter evolution at a later date. Since the concept of a higher evolvement of quality (qualigen) is foreign to the teacher, he carries out his duties unnaturalesquely and therefore incorrectly. In the house of God works the vicar. To a greater or lesser extent he prepares the evolutionally younger – the older generation – for death, that sublime process of transformation. He too does this unnaturalesquely and equally wrongly. Both teacher and priest are servants of a large-scale operation which can only exist as long as a shortage of food and spiritual and mental torpor prevail. All of which are prerequisites for a traffic in body and soul. The reason that the subtlety of this abysmal iniquity is so hard to apprehend is because at school and in church the human factory-fodder are told that this state of affairs has been ordained by God.

Neither of these two, the teacher just as little as the spiritual guide, are able to explain the sense and purpose of existence and the formation of the body. Both stolidly adhere to the decrees of Old and New Testaments and believe, God only knows how, that they have thereby fulfilled their ritual obligations. For this reason it is high time that the meaning and purpose of life should be explained plainly and clearly. After a person's life-energies (which gave them a sense of direction, a right-side-upness and enabled them to overcome gravity) have been extinguished, it is worth considering how such a lump of human flesh is transformed into spirit. This is analogous to the conversion of a crab-apple or a perry-pear into alcohol. In other words: how, after its demise, is subjective intellectuality (or the subordinate mental functions) of the body returning to the Earth reconstituted and developed into metaphysical products of fermentation with the aid of the Earth's cycloid motion?

The first thing that comes into being in this process is the so-called ground-climate. In other words, the geospheric frequency or soil-potential from which atmospheric frequency or mountain climate subsequently evolves. In these two metaphysical, and therefore more highly energetic intermediate zones of development, a dynamic, equilibrating activity takes place, which here will be referred to as the 'plasmolytic interaction of pressure and suction elements'. This interactive activity, which continuously triggers off processes in which crude and quality matter are reconstituted and built up, will be described in more detail later on.

More importantly, however, this interaction between threshold-magnitudes[83] can actually be copied with the use of specially-shaped devices. These function according to natural principles and at the same time ensure the almost totally controllable replenishment of foodstuffs. Moreover, the interaction also provides an almost-free source of power, thereby making this purely creative work a delight instead of a duty.

This has little or nothing to do with a discussion of natural philosophy, but with the examination of viable and practical possibilities for regulating the processes of growth, for restoring almost totally independent livelihoods, and for manufacturing biomachines. These biomachines provide the basis for generating near cost-free power and for producing almost total freedom of movement – the ability to travel over land, in the air and on or under water. In addition it concerns the designs of stable, reliable machines for generating a certain form of electricity. It should also be mentioned in passing that almost any desired increase and qualitative improvement of metals can also be achieved in this way.

In all seriousness this concerns the regulation of everything we see around us or are otherwise aware of. It concerns the universal promotion of growth and the faithful copying of all the products of growth that humanity requires for comfort and the maintenance of life, each according to their individual needs. At first sight this may all seem pure fantasy or the product of a warped mind. However, it will be seen to be just as down-to-earth and realistic as it now appears fantastic, once the physical and metaphysical processes involved have been elucidated and understood. People whose reasoning is deductive will always have difficulty in comprehension, because every dynamic, metaphysical phenomenon (in this case thoughts) leaves a physical imprint of its developmental

[83] 'Threshold magnitudes': Substances at the extreme limit of matter, or quasi-immaterial substances at the inter-etheric boundary between matter and pure energy. – Ed.

pathway, and therefore the convolutions of their brains become deformed. For the same reason even a healthy thinker will often be thrown off course and so the infinite extent and wave-like form of his thought cannot reach its ultimate conclusion. The thought then becomes offensive, is rejected and as a result becomes solidified and laid out cold through the nascent development of heat.

Growth and Seed Production as the Natural Goal of Evolution

Every movement has its beginning and its goal. The goal in turn is the beginning of an end without end. A movement without a goal is senseless. Were there no new beginning once this goal was reached, however, death and solidification would result. This is especially applicable to understanding 'plasmolytic motion', which is a prerequisite if we are to define 'growth'. To put it briefly, *growth* is the product of solidified, energetic qualigen with insufficient intrinsic potential. It is no longer able to co-act in higher levels of re-combination and formation, and solidifies under the influence of light and the Sun's heat. Just as warmed-up by-products of thinking calcify, so too does the qualigen, formerly in an energetic state, solidify into products of growth.

Naturopaths call the path of evolution the *spiral of life*, which is referred to here as the '*cycloid space-curve*'. In this cycloidal life-evolving path, pressure and suction forces develop similar to those that we know of in sexual responses. In its forward thrust, dynagen projects essences with an intrinsic spacial and potential vacuity ahead of it, like rockets. This makes it look as though the fluid seminal matter is expelled, whereas it is actually *sucked in*. What actually happens is that in its invisible, forward surge, the dynagen loses high-grade, spacially vacuous and potential-deficient essences in the process. It ejects ballast as it were, in order to increase its own speed and penetrating power. It is this energy and movement that forms and builds up the blood. The only difference is that instead of a straight advance, here we are concerned with a rotary biocurrent, which gyrates about its own axis. Due to the accretion of these completely solidified, deficient essences on the inner wall-surfaces of blood vessels the same kind of double-spiral motion evolves that we are still able to see in natural, undisturbed streams and rivers. If such conducting vessels are copied naturalesquely, then any water flowing through them will become cool, fresh, dynamic and free of gas. *A highly potent concentration of energy comes into being.*

If liquid or gaseous organisms (water or air) are swirled *cycloidally* in a high-speed centripulser,[84] the ensuing build-up of dynagen or qualigen

---

[84] 'Centripulser': Is a rotating device producing the sequential compaction (pressure) and rarefaction (suction) of the medium in question (air or water). In this process centripetence and centrifugence are active simultaneously, but centripetence predominates. It is described in *The Energy Revolution* vol.4 of the Ecotechnology series. – Ed.

levitates with such tremendous force that it also lifts the device that produce it in its wake. Taken to its naturalesque conclusion, this concept would result in the design of an ideal aircraft that readily ascends with hardly any fuel, or an ideal submarine which is lifted by the keel (by the seat of its pants) and dives just as easily. All this with hardly any propellent, because this system of motion works not under duress, but with *desire*.

The 'plasmolytic force' is triggered by an orbital outward-and-inward movement which originates at the periphery and acts along the longitudinal axis. It is therefore a question of spacial forces, because they permeate and fill the whole space. The product is born out of a screw-form motion in which pressure and suction evolve and operate on a common axis.

'Plasmolytic' motion is a seed-producing motion and springs forth from a highly excited protoplasm. In this process the physical production of seed is initiated through a longitudinal (up-and-down) stimulating motion, where-as a transverse, stimulating motion is active producing a metaphysical out-birth of energy. In this respect the type of motion originating from the material axis and the explosive force acting from the inside outwards are disintegrative and destructive. On the other hand the force acting from the outside inwards is creative and life-promoting.

Formative and Destructive Syntheses

The enormous difference between technology and ecotechnology will now be explained for the first time. Technology initiates motive impulses that destroy the substances required for growth and development. Eco-technology on the other hand, triggers a concentration of growth-enhancing substances. In other words, today's technology, physics, chemistry, *etc.*, gen-erate impulses that create products harmful to reproduction, regeneration and growth. Ecotechnology, ecophysics and ecochemistry generate impulses that create products beneficial to growth and reproduction.

Both technologist and ecotechnologist trigger only the energising impulse itself. They merely provide the impetus for the development of one effect or another. The technologist gives a *heating impulse*, the ecotechnologist a *cooling impulse*. The ecotechnologist is constructive; he provides for the uplifting advancement of evolution and development. The technologist, however, gives the very stuff of evolution a kick, which sends it reeling uncontrollably back-wards and down to the very lowest level of evolution.[85] The technologist, chemist and physicist are therefore destroyers of evolution and the ecotechnologist, eco-physicist and ecochemist are the promoters of evolution. The ecotechnologist

---

[85] Fire produces ashes, which are no longer organic. While still possessed of a small spark of life, ashes are inorganic basic elements, which have to be built up into organic structures again from scratch. – Ed.

can achieve economic growth almost free of capital input, whereas the technologist produces an inexorable decline, an economic catastrophe, at great expense.

A high-grade synthesis takes place when geospheric ethericities, which are free, highly stimulated, uni-polar fructigenic elements (the chemist classifies these as belonging to the carbon group), consume or bind inward, counter-falling (gravitating) seminal substances (principally oxygen). This results in the formation of a negatively-charged ground-climate, which is a high-frequency, horizontally-propagating potential. It is the true foundation for the build-up of evolution-enhancing growth products and is known as the seed or germinal zone. It is an interstitial zone of development lying between the positively-charged atmosphere and the negatively-charged geosphere, and encompasses the whole depth of the root-zone.

The germinal zone can be so enriched mechanically that growth and formation automatically take place as if on a conveyor belt. All that is needed is the smallest impulse in order to produce a boundless profusion of reproductive and upwardly-evolving out-births. Contemporary technologists, physicists and chemists do precisely the opposite. With an enormous expenditure on wasteful preparation, space and material (petrol refineries, for example), they inaugurate impulses leading to reversed syntheses. Here the seminal substances bind or consume the true formative substances, the fructigens (oxidation is the combustion by seminal oxygen of fructigens, such as carbones in the form of petrol, oil).

With water, for example, every rise in temperature reduces its quality and condition, because the wrong form of oxidation is activated. In regulated waterways, most of which lack protective bank vegetation, millions of valuable energies are lost, which ought to be employed in building up the vegetation.

The technologist is thus a parasite or the Devil who destroys Creation. The ecotechnologist on the other hand, furthers the will of Creation. Through his or her activities all shortages in the necessities of life and all drudgery unworthy of human beings will cease almost instantly. Everyone will become independent and free, and will have the time to devote themselves exclusively to the environmental activity that corresponds to their natural abilities and particular interest. In this fashion every more or less talented person will become an artist in their own particular field of endeavour and through the general improvement in everything they produce, the general standard of living will also rise.

Matter and the movement of qualigen arising from it will be under the full control of future human beings, who will become the highest servants and at the same time the masters of Nature.[86] Fabulous harvests will provide them with the highest quality food and at the same time they will enjoy the almost absolute freedom of movement on land, in water and in the air, in recognition of the services they have rendered Creation and therefore God. With this, all struggle

---

[86] *"We cannot command Nature except by obeying her."* Sir Francis Bacon. – Ed.

for survival, for social equality, and for existence itself, indeed all wars over food and raw materials will automatically come to an end. By activating the right impulse, peace will be assured throughout the world, because no-one will need to take bread away from another. Whoever helps another is never his enemy.

The fundamental error of technology lies in its exploitation of the direct action of pressure and suction, which function axially→radially, centrifugally from inside outwards. Wise Nature on the other hand employs exactly the opposite motive forces, which always operate radially→axially, centripetally from outside inwards. By inaugurating the latter motion any heating and cooling effect can be produced and therefore living spaces heated or cooled naturally. People will be able to raise or lower water at will and all contemporary machines of whatever description and for whatever purpose will disappear without trace. The next generation will already be able to look back in horror at the age when the evolutionally younger (their parents and grandparents) made use of mechanical monstrosities that engulfed the whole world in unease, anxiety, trouble and worry.

Even the medical profession will have to change fundamentally. What Paracelsus foresaw will become reality: *"One day there will be a specific, which will suffocate every germ of disease at its inception. Humanity will know no sickness and will therefore become full of the joys of life."* Space and land will be available in sufficient measure to those who through their very existence will enable the build-up of raw materials throughout the whole evolutionary chain to occur. Deserts will be made to bloom again. Where milk and honey once flowed, we humans created wastelands through our senseless greed, whence all water fled. As the source of life it could neither reproduce, regenerate or upwardly evolve itself, nor could it increase and improve in quality. As a result it developed into the most dangerous legacy of human endeavour. The fields became barren and the noblest high-forest, the natural donor of fresh blood, slowly, but surely began to die.

False and Faulty Management

Today's engineers, physicists, chemists and exponents of science had a bad education and were inculcated with the axioms, principles and dogmas of speculative thinkers and intuitionless individuals. They have all become servants of a decomposing and analysing principle, the principle of dissolution. They were forced to try first one thing and then another with a groping millimetre by millimetre strategy; they made no advance worthy of the name, but only went backwards. Because of their direct approach, the severest harm to the environment was inevitable. Wise Nature takes the very opposite road. Her evolutionary paths and processes are always *indirect*. This perception alone is enough to bring the mundane edifice of science crashing down. Religious instruction was no less of a hazard, because it perverted or perhaps even intentionally concealed the real reason for the creation of the body.

Technology is derived from ancient Greek prefix *techno*, which, amongst other things, can be translated as self-delusion or self-deception. Nothing further need be said! What now follows is therefore only a pointer as to how things could be done more naturalesquely. This is how one day they will have to be done if the present situation is ever to be redeemed and reversed. Once it becomes clear to all humanity that they are victims of the present situation, then the whole charade will be over.

The Mechanical Equivalent of Heat

One of the most dangerous fallacies, if not actually an intentional deception, is the Law of Conservation of Energy and the Law of the Mechanical Equivalent of Heat propounded by a certain Robert Mayer. The build-up of heat always takes place at the expense of vitality and heat can only reach excessive levels if retroactive syntheses occur, or when for any reason the seminal matter, oxygen, becomes free, uni-polar and thereby aggressive. On the other hand, heat makes fructigens (carbones) inactive and more easily bound. Every increase in heat occurs at the expense of the development of life. In order to heat up $1m^3$ of water by $0.1°C$ an input of energy equal to 42,700 kilogram-metres (kgm) is required. From this it can perhaps be appreciated just how much energy is lost if it is not bound in processes of organic synthesis.

Wherever the force of levitation is missing, then the force of gravity dominates. Where gravity alone prevails, then the momentum is also absent to which the naturalesque impulse or mechanical movement gives rise. In this respect, the force of gravity is only necessary and interesting for initiating the formation of the developmentally-important form of energy. The maintenance of the impulse therefore signifies the maintenance of formative life-force, for which the presence of a safety valve, the 'brake', is imperative. For this reason excessive heat represents Nature's will to *eliminate*, to which everything falls victim that is diseased or has become unusable for further evolution. The best symbol of this process is Hell, where sinning souls, those unfit for life, are allegedly condemned to burn for eternity.

The fate of life itself is a question of the way it evolves. This in turn depends on how the *two types of synthesis* are organised, both of which can be regulated through cycloid motion once the given medium has been dosed with the appropriate elements and ingredients. It therefore lies in humanity's power to produce the desired impulse. In this way the formation and decomposition of blood can be so ordered that (a) only the latency is actuated, suited to processes of higher transformation and growth, or (b) whatever is inappropriate for the cultivable path of evolution, for one reason or another, is made to repeat the whole process from the beginning.

Knowledge and Science

Western philosophy doubtlessly obtained its most powerful impetus from physics which, although founded on basic truths, only supported itself on *effects*, never on causes. With causes *faits* can be *accomplis*, once the process of growth and transformation is known through which physical actualities are created. The reason why these cannot arise naturalesquely today is due to the use of the wrong physical (mechanical) motion. Goethe was severely critical of science because it disregarded the inner essence, the philosopher's stone, as it were, of the thing itself around which everything revolves. To which Goethe commented, *"By this one may recognise a man of learning, for what he cannot touch lies miles beyond his understanding"*. This inner nucleus can be developed in one way or another according to the type and direction of rotation.

All genius springs from an artistic fantasy, whereas a moribund fantasy allows no creative pennies to drop. An impulse is energy *in statu nascendi*. It can serve either integration (growth) or disintegration (decay). Cause and effect are the result of cycloid space-curve motion from which life or death springs forth. The skill merely resides in the organisation of the rhythm of both types of synthesis. Both must always be active and at all times present, and must make use of what is useful and harm what is harmful. Fantasy therefore is the unleashing of a fantastic impulse.

Naturally it would take more than one blow to fell the tree of unnaturalesque knowledge because it is so rooted in all branches of industry. Amidst the ensuing turmoil therefore, the difference between chalk and cheese would be hard to distinguish. The present order would also be upset in one fell swoop if its underpinning financial markets were dealt with carelessly. Should no suitable precautions be taken in this area then the whole financial and commercial applecart would be thrown into total disarray and a chaos of unimaginable proportions would ensue.

The last war [WW2] was a good teacher and led to the realisation, so aptly expressed by Goethe, *"Now here I stand, a fool so poor, and just as clever as before"*. The war and the post-war period have enabled the after-effects of a dead science to be experienced so quickly that causes and effects can be perceived by a single generation. Previously it was only later generations who began to feel the effects of science's misguided activity, but they could not discern the causes. Ultimately science and the war provided the motivation to investigate causes, because the people themselves experienced the effects in quick succession.

Naturalesque Activity

The *modus operandi* arising from the activation of '*plasmolytic essences*' of pressure and suction will sweep away (or suck in its wake) all those whose minds were too stolid or too stupid to realise that what benefits one benefits another. A love of one's neighbour will evolve out of sheer self-contentment. The mysterious *ur*-force that Goethe called the 'Eternally Female' must also be divested of its enigmatic veil. It is equally necessary that the 'stronger' male sex be made aware of its own weaknesses and thereby become more careful. We should not blindly believe everything and allow ourselves to be governed by laws that do not exist, for all that prevails in this evolutionary happening is *rhythm*. This rhythm strengthens whatever lives naturalesquely and moves itself in such a way that apart from its muscular strength, the power of its mind also grows through mental exercise. Spiritual sport is just as important as physical sport, the former being understood as a life-affirming activity, whose principal function is to foster the care of that fount of qualigen, the physically first-born – water.

This source of qualigen, so vital to the survival of all fields of human endeavour, cannot be explained scientifically. It defies all description, because so much is happening or has already happened in this colourless, tasteless, odourless and formless water that, quite literally, all that is needed is just a small impulse to initiate something good, bad or indifferent. The fact that no major catastrophe has so far happened is due to the sensitivity of key threshold-elements contained in all water. These critical threshold-elements are in a state of equilibrium. They can move neither forwards no backwards. As intermediary bacteriophagous essences they stand between life and death. It is due to this neutral condition that anything has been able to continue to grow at all. Even exact science was unable to poke its nose in everywhere, and wise Nature has been able to outwit its own experts.

Since patents applications have been made, which definitively clarify the internal processes in water and air, no further details can be given here. Due to this patent-related impediment, however, the problem of naturalesque methods of working will solve itself. Everybody will research in depth out of primal self-interest. Digging deeply, they will become composters instead of impostors, thus bringing every ecclesiastical potentate and all intrigue to a sudden end.

Goethe as Biologist

It is difficult for an engineer, a physicist or a chemist to construe the poetry of *Faust* and Goethe's observations of Nature as the best form of biology. However, if the purpose of life is deemed to consist of making a small piece of this Earth productive, then the bio-ecological compass of this German poet-prince's persuasive intuition will be perfectly understood. To make a small plot of ground fertile in effect means nothing more than to set a very small and insignificant threshold-element in motion, which possesses just enough intrinsic potential to overcome its own body-weight, but not enough to levitate. All that it lacks, therefore, is the living spark in order to trigger its practical cultivation.

Cycloid motion, through which plasmolytic, germinal entities come into being, is all that is required to make it practically possible. These immediately distribute themselves vertically and increase the quantity of the next higher product of evolution and qualitatively improve it. A forest of new trees of knowledge springs up from this realisation, of which only two will be addressed briefly here. These two trees of knowledge will suffice to uproot today's science and technology. If the following chapter does not jolt them out of their complacency, then nothing will.

The Creation of Impulsion and Expulsion through Cycloid Motion

It is common knowledge that all life comes into being through movement. Since everything is bipolar in nature, there are also two different types of motion. One of these leads directly to heaven and the other to hell. Between them lies a biological state of utter indifference, a purifying purgatory, a labile state of equilibrium at all levels of life and development.

The great theologians were therefore quite excellent biologists, who celebrated the circumstances of transfiguration in the Holy Mass so admirably that no great skill is required to draw very practical conclusions from such a doctrine. Since pressure and suction operate simultaneously on a common biological axis, all that need be done is merely to transfer the ritual of the Mass into rational practice in order, for example, to atomise water into air. In such a way an ideal means for producing mechanical pressure is born, which is significantly better and above all, more naturalesque than all contemporary means of pressure and explosion put together.

All water contains bipolar gases. It is a physical frontier zone, so to speak, in which all sorts of interesting bacteriophagous, intermediary elements of bipolar nature are contained. These *philosopher's stones* are almost impossible to detect. If quite ordinary water, with a high gas content and deficient in energy, is swirled in cycloid centripulsers, its whole condition can be completely reversed, producing a deficiency of gas and a wealth of energy. This happens as a result of the repetitive generation and development of plasmolytic pressure and suction in the centripulser. A physical threshold state develops requiring only a small heat-intensifying pressure (a combination of mechanical and physical effects) to transform this atomised critical mass into its immediately higher, aeriform state. To bring this about atmospheric oxygen has to be introduced through needle jets.

A new air with very high potential (frequency or vibration) is instantly produced in this way. So that the whole thing does not shoot straight up into Nirvana, it simultaneously gives birth to a multiplicity of higher carrier substances in order to produce the necessary counterweighing pressure. This restraining force helps the mass of the Earth to counteract the force of levitation. The force of gravity is therefore a concept to be taken very relatively. It becomes utterly and completely uninteresting if one merely moves a mass of very ordinary air cycloidally. With the use of this ultimate fuel the scales are tipped, the last counterweight falls away and there is no longer any excuse for upholding such cumbersome concepts. There is no longer any *"she half-pulled him, he half-sank on her"*.[87] All that is left is the purest levitation. For this reason this autarchical upward propulsion must be handled with the very greatest care.

But let us return to the naturalesque inauguration of this silent, odourless, expansive force. Actually it is only required to start cycloid motion and only initiates the overture, as it were. It is therefore merely a practical aid for unleashing the *ur*-force.

Physics and Metaphysics

It was already stressed at the beginning that there are no technical or scientific expressions for metaphysics. Since no exact terminology exists, complicated paraphrasing is required to describe the higher origins and causes of motion and formation. Physics makes use of various pistons, substances, mechanical or physical forces in order to create motion. Metaphysics uses motive forces that are similar in principle but are raised one octave higher. In every case, only those forces which improve the quality and condition of the substance to be moved are used by the metaphysicist, who knows that all natural processes are interconnected with physical (mechanical) motion.

These mechanical and physical forces, however, are applied only in the smallest amount in order to provide the necessary impulses to activate the more important plasmolytic forces. The metaphysicist utilises these higher and more powerful energies to initiate a new interchange of substances. In doing so he achieves a nascent kinetic energy, an inner potential, strong enough to interact with the surrounding concentrations of energy of opposite potential. Motive forces and velocities are produced of a magnitude contemporary physicists and engineers are incapable of conceiving, because their thinking is logical and not *bio*logical or *eco*logical. They employ direct

---

[87] Quotation from Goethe. – Ed.

methods and approaches which produce minimal effects or usable energy and inflict severe damage on growth and the environment generally.

Today's sophist has stamped the conception of the world with an alien character by disseminating the view that a certain quantity of energy always remains constant and against which an immutable force of gravity is active. If today this gravitational force is hardly ever mentioned and attempts are made to maintain and intensify locomotive power, it is only indicative of the search for other and better solutions. They cannot be found, however, because thousands of years of deductive thinking have deformed our processes of thought. On this well-grooved track every flash of inspiration and its outcome has been and is derailed.

The human organism can be compared to a high-pressure boiler whose safety valve blows when the maximum permissible steam or gas pressure is exceeded, whereupon the boiler flashes back from high to medium pressure. This results in the strong accretion and deposition of boiler-scale so that eventually the boiler can only be used at medium pressure. Its efficiency constantly decreases as the accumulation of scale increases. Metaphysically speaking, this means that with the use of naturalesque and potential-enhancing processes, not only can high-quality solidification of basic elements (growth in the form of food) be brought about, but also quantitatively and qualitatively increased growth of minerals and metals too. In addition, a waxing, formative, motive power is produced, which will make problems of food supply, raw materials and energy a thing of the past.

Through the naturalesque creation of potential and the correct method of intensifying the causal processes of substance interchange and potential alternation, all waste and reversionary matter are always precipitated out in a *more evolved state*. The whole secret of evolution therefore lies in the self-regulating increase of impulsion and expulsion that takes place under normal conditions, which is inherent in the rhythm of life and its evolutionary purpose.

Today's scientists unquestioningly adopted rigid theorems and principles and turned them into dogmas. These were foisted on them by a group of people who had no desire to increase and improve the quality of foods and other necessities of life. Any increase in mental ability was of absolutely no interest to them because it did not serve their plans for domination. The effect of this was a constant increase in labour itself and the number of people forced to work. More and more these dubious individuals sabotaged the mutually-amplifying interaction between impulsion and expulsion. In the high-pressure boiler of the Earth the expansive pressure dropped due to a reduction in the ennoblement of impulsive and levitative substances. At the same time the quality and quantity of precipitated reversionary matter, the amount of boiler-scale, constantly diminished and was of less and less value. In other words, fewer and fewer compounds of mineral, metallic or vegetable raw materials (growth products) were produced, and all the while the quality of these products of solidification declined. As was said earlier, these products are the basic elements for the development of higher potential.

The bio-ecological outcome of this was a shortage of food coupled with a catastrophic decline in mental health and vigour. This opened the way to a thriving traffic in the necessities of life founded on the absence of high-quality precipitates and a dearth of powerful intuitive abilities or products of the mind. The poor wretches of labouring human oxen were ultimately led to believe that the wages they received for their toil and trouble would be paid in heaven.

The Most Iniquitous Deception

Philo was the founder of philosophy. Classical physics is none other than applied philosophy. Only through the most disastrous experiences will people become wiser and renounce this form of science.

The question *"Why does the heart beat?"* irritates every doctor. The question *"What actually is motion?"* reduces every mechanician or professor of physics to a state of fury, particularly if he is asked, *"What is to be understood by mechanical or physical motion and what differences exist between them?"*. Neither of them are able to give an answer. If they were aware of the differences, they would also know what *motion* actually signifies and would need to learn neither physics nor mechanics in order to teach it. They would also have no need of their professions once the question *"what actually is motion?"* has been thoroughly clarified.

Motion is the product of potential differences of physical or metaphysical origin. In this regard it is of crucial importance whether the motion is initiated from outside-inwards or from inside-outwards. Apart from these, further differences also exist, however. A stone, for example, can be moved physically by an external agency, with a shove or a kick, or it can be moved metaphysically by being made to leap up musically or through the agency of music. This is what happened to the walls of Jericho, which were demolished by the impact of the sound of trumpets. Today this can no longer be explained, because the motional rhythm or the previously discussed mysterious process of insuction or inner attraction, engendered by an external impulse, has been ignored. This is how vibration begins and explains in a flash the concept of *motion*. Thus the whole of science can be literally reduced to its own atoms and a new science of life created from the analysed remains.

This knowledge has been so thoroughly buried that today nobody can say what *life* actually is, whence it comes and where it goes. No-one any longer knows that life or autonomous movement is the product of a *vibrating vibration* and a *moving movement*. This rhythmically-moving rhythm arises when the two components of pressure and suction, or the 'backward-and-forward', the 'up-and-down' and the 'inward-and-outward', are rotated to the left and to the right in three-quarter time about a common developmental axis. Here positive potency is to be regarded as pressure and the negative potency as suction, or the former as the gravitating and the latter as the levitating. Expressed differently, the attracting and the repelling must be moved in opposite directions along the axis. Without a model to demonstrate this practically, it is almost impossible to picture what form this motion takes. The product of this primeval, elemental motion is the 'obverse', the indifferent, formless, bacteriophagous mist that forms between existence and non-existence, which expressed differently is synonymous with 'Life'.

All the philosophasters (the be-wigged and be-gowned) had to do, therefore, was to suppress the natural communion between the basic elements of evolution through the enforced application of classical physics and the doctrines of mechanics. Ecologically and biologically this resulted in the elimination of sources of qualigen, which inevitably led to a worsening shortage of food and spiritual impoverishment. Thus reduced to the lowest level of existence, the last vestiges of human dignity were in great danger of being trampled underfoot by collectivism (communism). It is therefore high time to set about reversing this process.

The Birth of Basic Elements and Subtle Matter

All basic matter carries within it the seed (DNA) of reproduction and upward evolution (for the increase and qualitative improvement of higher forms of matter). The sole purpose of the force of physical weight, the force of gravity, is to initiate inner processes of substance interchange and alternation of potential. These encompass two fundamentally different orders of metaphysical dynamics which are:

(1) groups of essences producing metaphysical pressure or repulsion;
(2) groups of essences producing metaphysical suction or attraction.

If, as was explained above, the retaining bonds of high-frequency water are removed, or it is atomically decompressed or atomised, then an increased and intensified, more highly-charged build-up of air-pressure normally results. This in turn serves for the inward fall of seminal substances. If this expansive force encounters a resistance in the form of a piston, then the piston is moved by elemental forces, resulting in a natural and almost silent motive impulse. The pressure-release valve in this instance is the mechanical piston. If the potential and polarity of high-frequency water, super-saturated with metal embryos or mineral structures (carbones), is reduced by diffuse solar energies (oxygen) under the hermetic exclusion of air and with slight warming, then a convergent, centre-directed heat-pressure evolves. At the same time, however, an inner de-energising of highly active fructigenic elements, the matrices of ions or electrons, takes place, resulting in their solidification. A growth in minerals and metals ensues, whereby the weight of the parent materials does not reduce, but increases in the same way that the girth and weight of a female creature normally does after conception. This is because every conception creates the preconditions for further conception – the first division of a cell, for example, creates the conditions for the later division of the two original halves.

If high-frequency water is not de-energised or its energy modified, if the motion responsible for the generation of high-frequency continues unabated, then the horizontal propagation of negatively-charged embryons[88] takes place, i.e. the emission of negatively charged, ion-like or electron-like embryonic formations. The dynamic aspect of the geosphere then experiences a growth, which expands parallel to the ground-surface and has the same wave-like crests and troughs as the ground-surface itself, which should be viewed as solidified geospheric energy. It is that part of the terrestrial mass, whose upper layers are sealed off by the film of protoplasms, the virgin hymen of the birth-giving Earth.

If these highly potent, wavelike formations are bombarded by diffuse, lightless and all-penetrating peak-energies of seminal matter, the Sun's (cathodic) 'cosmic' rays, then a negatively super-charged build-up of groundwater results. The groundwater itself is a by-product of synthesis which continuously binds all newly produced fructigenic structures. The effect of this is to cause an unceasing rise and fall of the groundwater body. Because groundwater constantly discharges its accumulated energy into the root-tips of the vegetation, its potential builds up and discharges at rhythmical intervals. This gives rise to the pulsation of the Blood of the Earth.

The discharge from the high-frequency accumulator (the groundwater) into the root-tips (root-protoplasms) in turn encounters the discharges from the cathode system of the Sun. The outcome of this interaction is then organic growth. or expressed otherwise, the self-concentrating precipitation of frequency-depleted matter with a lower level of dynamic energy These agglomerations contain all the seminal, fructigenic and stimulating elements required for the creation of an organic whole. That is, apart from the principal basic elements, they also contain catalytic motion and interaction inciters. These provide the true impulse enabling reproduction and ennoblement to reach their perfection on the path of evolution.

That this should actually happen requires the prior existence of the appropriate preconditions; the proper proportions between various basic elements, the exclusion of atmospheric influences and the presence of the necessary resistances. Through these the heat-inducing elements are de-activated, leading to the evolution of cooling influences vital to growth. Nature

[88] 'Embryons': Are the energetic counterpart of physical embryos. – Ed.

builds up this *coolth* through the consumption of seminal substances (oxygen). In this large refrigerator the foolish men-folk will be swallowed up, thereby fulfilling their appointed destiny.

The Group of Metaphysical Pressural Elements

This encompasses three possibilities.

1. The virtually cost-free impulse-imparting motion of Repulsators and Repulsines (substance-interchanging and potential-alternating machines);
2. The controlled increase of the growth of catalysts and the growth of high-grade metals and minerals;
3. The increase and ennoblement of organic growth.

This metaphysical pressure-element group therefore enables the super-abundance of food to be regulated almost at will. Its other associated fields of application are:

- the creation and control of temperature;
- the production of heat and ice;
- the ecotechnical raising of water;
- the generation of motive power for all types of machine;
- the propulsion system for vessels that have to operate in the absence of air, such as submarines.

The Group of Metaphysical Tractive or Suctional Elements

If ordinary air is moved *cycloidally*, it can be just as easily enriched as can ordinary water in order to increase the performance of metaphysical products. Then, after indifferent substances, which are contained in the air in suspension, have first been dissociated, products are formed that are the result of primary synthesis. In the process positive atmospheric potencies accumulate and are then transformed into negative dynagen structures, which are virtually carrier-less. Discarding all coarse, physical matter (atmospheric water particles), they rise autonomously as exponents of pure levitation. All motion-impeding resistances, even the aircraft itself in which this effect is produced, will be drawn up in their wake, because the *organic vacuum* which develops in front of the craft (a spaceless and formless state of quality and an almost total vacuity) has an *adsuctional function*. Whereas the air-masses (discarded matter) expelled sideways and towards the rear can only slowly achieve equilibrium with those that surround them. In this way a wedging pressure and hence a forward pressure is exerted on the teardrop-shaped body of an aircraft or submarine. This pressure acts along the longitudinal axis in the same direction as the suction (see fig. 9). The trout demonstrates this to us in every natural, untouched watercourse: its arrowlike flight upstream against the current making the faces of our physicists grow longer and longer.

Steerage is achieved with an articulated swivelling Repulsine-head, which creates an organic vacuum in the desired direction of travel. This ideal aircraft therefore only requires fin-like stabilising surfaces (similar to flying fish) and somewhat larger tail-fins in order to guide and move it through the air, or in the case of submarines, in and under water at lighting speed.

The operating range is unlimited, because the fuel (which is also growth-promoting) required to provide the impulse for cycloid or plasmolytic motion can be generated from the surrounding media of water and air in unlimited quantity, through inbuilt Repulsators (high-frequency water-generators).

The same supporting force (floating power) that also carries this dung-heap Earth and rotates it about its own axis can therefore be generated almost without effort and expense. Orientation and control is done by the pilot, who also owes his or her own orientational faculties and physical strength to the

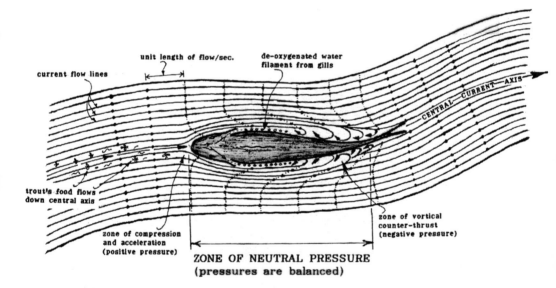

Fig. 9. The stationary trout.[89] The trout normally swims in the middle of the central current vortex, where the water is densest and coldest. Due to the volume of its body, the individual current filaments are displaced and compressed. This causes their acceleration and eventually their critical velocities are exceeded, which results in the formation of vortices (counter-currents) along the rear part of the body. These vortices act counter to the direction of the current and provide the counter-thrust required by the trout to remain stationary in this fast flowing water. If it needs to accelerate upstream, then it flaps its gills, creating a further vortex train along its flanks, thus increasing the counter-thrust upstream. The more rapid the gill-movements, the faster the trout moves upwards against the current.

[89] This illustration has been reproduced from the editor's own book, *Living Energies*, for better understanding. – Ed.

cycloid movement of the blood, whence the supplementary forces arise that carry, shift and orientate him or her, and sustain life.

This is how the general interaction takes place in which every creature is reunited with its metaphysical, genetic stock. In this way everything serves the will of the Creator on a common evolutionary axis. Everything furthers the general purposes of reproduction, regeneration, upward development, evolution, growth and dissolution. Everything also provides for the structuring and build-up of crude and refined matter from which the cultivable *mobile vitalis* is produced. The disintegrative forces are also intensified, which, for one reason or another, dispose of those things that have failed to meet the requirements of the path of evolution.

It is precisely this *disintegrative force* that today's science is escalating and with which it brakes the cultivable path of evolution. It brings about an evolutionary standstill and inaugurates the collapse of western culture. *The influences of fire and heat belong to the most dangerous, retrogressive products of synthesis.*

A General Explanation of 'Plasmolytic Motion'

In order to get some idea of the formation of blood and sap, the water-processing systems of plants should be studied. The contribution of plants to the amount of water accumulated in the atmosphere is demonstrated by the fact that a large birch tree contributes about 7,000 litres (1,540 gals) of water over the course of a single summer, representing the precipitated waste-material accruing from the formation of sap. This volume of water is transpired into the surrounding air via the leaves. Stated more accurately: the primary products of synthesis carry up this quantity of water with them in their wake in the smallest packets and hand them on to the levitational substances of the air.

Nature has equipped every plant with its water-production system and every plant produces its own kind of water. The *primary* products of synthesis infuse themselves into such water. In a certain sense they carry the abode for their brief future sojourn in the atmosphere about with them. This explains how positively-charged high-frequency essences are born from the negatively-charged atmosphere, which together with heavy, solidified Earth-masses counteract the levitational forces that develop in higher regions.

In some situations it so happens that these extraordinarily powerful, primary products of synthesis (levitational forces) have not been sufficiently bound by the counteractive influences of light and heat. In barren high-mountain regions they are insufficiently encumbered with ballasting material. In these locations the phenomena of so-called ball-lightning, giant-size Elmo's fire in egg-shaped form, flares up, which in most instances can be observed above the very tops of the trees towards midday. They sizzle like lighted baptismal candles and then quickly go out. The cloud-like condensation trails they produce, akin to those formed by aircraft in tight turns, are the surest sign that there will be a heavy thunderstorm a few hours later.

At certain locations high up in the mountains and particularly out of crevasses in glaciers, very sudden vortex-like up-winds burst forth, which swirl large quantities of snow upwards with tremendous force. Even chunks of ice are torn away and lifted high up into the air. This is a sign that the counterweighing seminal substances (the force of gravity plus oxygen) are unable to consume (bind) the soaring levitational forces. In the majority of such cases light intensity and heat are insufficient, which like warming water normally render energetic fructigens inactive. The reverting levitational substances then manifest themselves as sparkling bubbles of carbon-dioxide, primarily where the strongest incident light and heat are active. These reversionary products are the true threshold elements, which immediately reproduce and upwardly evolve when such water is moved cycloidally under the exclusion of light and heat. Under such conditions, for example, new water is created, which, floating freely, no longer obeys the law of gravity and rises up a vertical pipe despite the pressure of air weighing down upon it. Merely touching this bubble of water with the end of a metal needle is enough to discharge it, whereupon it falls back instantaneously.

This explains how atomically-heavy masses of oxygen, in combination with finely-dispersed large agglomerations of atmospheric water, are able to float. In these structures, though difficult to determine exactly, the Earth's high-frequency and primary products of synthesis are contained in the form of free energy. They also make springwater specifically-heavier, dense and cool. The water in high-mountain springs is full of levitative power, and inasmuch as high-grade springwater is actually present, the spring *springs upwards*. The peoples of ancient cultures made use of this process to lift enormous quantities of fresh water to any desired height. In this regard, the rising pipes must be shaped like blood-vessels, i.e. they must incorporate cycloid space-curves.

All these observations give rise to a true conception of the world, which revolves like a fertilised yolk-sac within the surrounding concentration of energetic fructigens. The same process of development takes place on a large scale that daily occurs in every incubating chicken's egg. Moreover, it is also the same process that can be observed in a natural, untouched stream. Along its longitudinal axis a zone of concentrated dynagen comes into being through the free and unimpeded movement of cycloidally-swirled fructigens, which react to biocentripetence. Seminal substances congregating along the flow-axis are surrounded by these fructigens and bound by them. This results in the formation of calyx-like or funnel-shaped, refluent energy flows. These in turn activate the marvellous twofold brake, which retards forward movement of the main body of water as it flows down a gradient.

In this vortex or flow-brake, a trout is able to float motionlessly. Whenever it intensifies the transformative processes with the aid of the cycloid curves of its gills, it is accelerated upstream at the speed of lightning. This is yet another clue as to how naturalesquely-built aircraft and submarines can also move with equal speed.

Every plant is an extremely efficient producer of water. On no account is it any kind of water-consumer. The root-tips of all species of plant are hermetically enclosed by protoplasmic structures, which even gaseous elements are unable to penetrate. The formation of the later structure with more highly evolved basic elements is only possible through this hygienic filtration of all physical raw material (solid, liquid and gaseous). This growth will be impaired, for example, if root-protoplasms are etched away by artificial fertiliser, or if they wither or become porous. Inferior substances are then up-taken and inaugurate cancerous decay. In trees this begins inside the base of the trunk and spreads, blackening the branches, which rot right through to the pith where they spring from the trunk. To an alert forester this is an indication that the proper mixture of species is out of balance and that the accumulation of geospheric soil-potential has been disturbed or is taking place unnaturalesquely.

Every cell creates its own protoplasmic nucleus autonomously. Every protoplasmic structure generates new protoplasmic energies for its own use. This explains, for example, how a large chestnut tree can effortlessly counteract the heavy weight of its leaves, flowers and nuts, which collectively weigh several tonnes. In such a tree-community millions of minute power-stations operate jointly in order to maintain the carrying capacity, tractive energies and orientational force, i.e. the life-force. This is the reason why sap supposedly rises in trees often a hundred metres high. This is none other than the consequence of a rhythm against which a deracinated humanity, imprisoned within walls of its own creation[90], toils by the sweat of its brow.

All visible phenomena, it is immaterial whether these relate to a solid, liquid or gaseous product of growth, or a species of water, blood or sap, are the reversionary products of primary products of synthesis. It is an outfall of energetically depleted matter which possesses too little intrinsic potential and solidifies under the influence of light and heat. Due to the cycloid motion of the Earth they are able to produce the physical seeds for their own reproduction and, as the finished products of raw materials, they die off as the plasmolytic forces are extinguished. Eventually falling back into the Earth, they enter a zone isolated from light, air and heat. There they are decomposed by analysing and vibrating energies in preparation for their contribution towards the build-up of the soil's metaphysical and energetic potential.

---

[90] Here the German word 'freigemauerten' is used. A 'Freimauerer' is a freemason and therefore in view of Viktor Schauberger's active disdain for the activities of the 'be-wigged and be-gowned', this phrase could also be interpreted as 'imprisoned by Freemasonry'. – Ed.

The dead bodies of human, animal and plant worlds do not decay to dust and ashes, but are the seeds that create the world of subtle matter and energy, from whose precipitates higher in-spire-ation stems. Through the agency of the ground-climate – the soil-potential – they enter the various organisms through the intake of nutrients and shape the formative, vital forces of the organism in question.

This process of growth and transformation, or the alternation of potential, which perpetually renews itself at any given moment in the ever-changing forms of manifestation, must be copied. In order to do this, protoplasma-like, naturalesquely-alloyed, pulsating devices are required, which must operate and rhythmically alternate at high speed. This will furnish proof that this structure-creating process is the one that furthers the natural course of evolution.

Thus the way has been opened for the Crown of Creation to become the highest servant of the Lord of this Creative Work and hence the director of a grandiose scheme of evolution. Perhaps the people of this century have been given a unique opportunity to become as Gods, as they struggle up this knife-edge ridge, fraught with the ever-present danger of plunging into the bottomless abyss. Whoever masters the process of metamorphosis in its formative sense acquires the abilities of the Creator. Whoever masters metamorphosis in its destructive sense becomes a tool and the servant of the devil, and it behoves such a person to carry out the work of destruction.

Organic Syntheses – Conclusion

A few examples will now demonstrate what energies are actually present in the various types of blood and sap. It should be noted in this regard that crude and subtle energies are not merely summated (increased arithmetically), but squared (increased geometrically) as well. It is only thus that the magnitude of elemental carrying and tractive forces becomes understandable. These can be controlled in miniature machines by regulating the rate of rotation.

As do all other plants, a grapevine cut off about 80cm (30in) above the ground exhibits symptoms of bleeding. In the process the sap is expressed with a pressure sufficient to raise a column of mercury in a barometer by 112mm. Desert plants generate plasmolytic pressural and suctional forces in the order of 100 atmospheres or more. We are not concerned here with the raising of a nutrient solution in the accepted sense, but with upward and downward pulsating, procreative events. With every pulsebeat the primary structures in the hierarchy of structures are produced. The higher and faster the beat, the higher the potential pressure evolving from the previous evolution of potential pressure. This ultimately reverses polarity into high frequency, which results in the creation of increasingly refined secondary products of synthesis. These reinforce every renewed process of growth and transformation so that out of the original precipitation into matter an instinctual awareness is born. An outflow arises out of this from which the mysterious possibility of overcoming one's own gravity or physical weight and its associated life-force spring forth.

This also explains how birds are able to fly, which they do with virtually no effort. At the bottom of the quill of every flight-feather there is a small motor in the form of an energy-sac (protoplasm), which with the slightest impulse-giving movement (beat of the wings) endows every particle of the wing with levitational substances. The bird does not fly, but it *is flown*. The fish too does not swim, but *is swum*. It steers itself in the desired direction with its fins, just as we do with our feet. It has no awareness of its own weight, just as a young, healthy human being experiences no oppressive weight on its feet. If the fish's supposedly air-filled swim-bladder or the bird's quill-sac is damaged, then despite fins and feathers, all swimming and flying are over. Even our own ability to move comes to an end if biovolts and bioampères can no longer be generated in the bloodstream of our own organism. The same applies to all our senses, which owe their activating force to the 'plasmolytic' motor that Leonardo da Vinci called *"il primo movere"*.

The impetus of Will is the waviform shaft about which this rotates, whose plan and elevation are identical. The working drawing for this we find in every well-formed hen's egg, the form of whose shell contains all the necessary constructional profiles.

Sven Hedin gives a very interesting and illustrative example in his book, *The Flight of the Great Horse*, where he describes aqueducts through which it was possible to grow the finest of cereals in the middle of Asia's waterless wastelands. These are the last remnants of an ancient people, who were still aware of the necessity to move the Blood of the Earth (water) cycloidally. These canals were laid between 1 and 2 metres underground and their gradients followed the contours of the ground surface. As a rule, therefore, their longitudinal shape is wave-form. These channels represent a sort of artificial geotherm (stratum of equal temperature). At intervals these conduits are interrupted by shafts which are inclined in the direction of flow. These are used for access, for the cleansing of the channel and to aid the water to regenerate itself with the induction of oxygen and carbon dioxide. The main orientation of these canals is north-south. However, where ground conditions permit, they are laid from west to east on principle (using the geostrophic effect on west→east flows). In such cases, and right in the middle of the desert, the vigorous growth of the noblest cereals is evident both to right and left of the underground canal. Wars either decimated those entrusted with the techniques of building these aqueducts or drove them from their homeland, and in this way the art of growing corn in desert regions was lost.

Similar installations were built by Moorish peoples. The only difference being that the Moors dug no canals. They dug funnel-shaped cisterns in the ground instead, which had definite longitudinal and transverse cross-sections. One or two relics of this aquaculture are still to be found in Lower Austria, where migrant monks tended their monastery gardens. As a rule these funnel-shaped cisterns were dug in marshy and sour ground, whereon a lush growth of sweet grasses appeared shortly thereafter.

Apart from a few scanty references in old chronicles and oral transmissions, more detailed information about these curious installations cannot be found. In this way the valuable knowledge of the essential nature of these water-cultures was lost. It was only observations in field and forest over many years that led to the rediscovery of this long-lost knowledge. In the near future its renaissance will change all forestry, agriculture and water resources management, for with the installation of such water conduits and cisterns it is possible to increase the productivity of the soil several times over.

These water conduits, or the chambers with unusual longitudinal and transverse cross-sections, have to be placed at a certain depth. They are none other than electrical devices, a special type of dynamo. They emit an animal radiation into the surrounding soil which is generated by the movement of water. This radiation is broadcast laterally and generates a stratified potential-field in the ground (in contrast to all surface canalisation, which discharges the ground). Following ground-contours, this potential-field is essentially magnetic in nature. It is propagated horizontally and in it centrifugating and centripetating pulsations develop. This regular rhythmical movement moves in towards the centre only to move outwards again.

Just as they did thousands of years ago, there are still 'initiates' today who watch over and see to it that the great secret of evolution is safeguarded from those whose interests are served by ensuring that humanity remains divorced and disconnected from Nature. It was not without good reason that Schiller told of the youth, who in his overwhelming desire for knowledge, ripped away the veil shrouding the image of truth. The priests found him the next day at the foot of the alter of Isis. Whatever he had seen or experienced his tongue was never able to express.

Human beings *are* what they *eat* and they will remain as animals for as long as the production of qualigen is suppressed. In this way a vicious circle is closed: polluted water can create no wholesome food; therefore contaminated water and poisoned food can create no healthy blood; therefore unhealthy blood inhibits a healthily functioning mind; therefore these degenerate qualities will then be passed on to later generations via the production of genetically impaired, seminal matter.

Western man is the victim of a life and energy-subverting principle. We have lost our greatest asset, our intuition, and have thereby become a creature even lower than the animals, a creature condemned to senseless toil. We will have to suffer the total loss of our bearings during our 1,000-year long period of re-evolvement. Nature has no use for toiling human beings. What she does need, however, is the extremely fine adjustment and regulation of the impulse with which all growth in Nature can be energised and regulated as circumstances demand. If this Will should be reawakened and the impulses magnified with the aid of ecotechnical appliances, then the effects of this period of decline will be overcome within a relatively short time.

Everything superior falls step by step upwards (levitates). Conversely, all that is inferior and unusable falls stage by stage downwards. The more useless such a thing is, the lower the level to which it has to fall until it reaches the place where the crucible for all unnatural products of synthesis glows darkly. It is here, where the dark fire smoulders, that everything is reduced to the lowest level of evolution by extremely aggressive seminal matter (oxygen). The sly priests called this place the hell of eternal damnation.

What they called heaven is in reality a concentration of energy similar to a fructigen-sac in a fertilised chicken's egg, which floats and slowly turns about its own axis under the influence of constantly-fluctuating variations in temperature. In the process, a gentle life-current will be emitted in the form of impulses which, with a moist tongue, can be sensed as a coolness at the pointed end of the egg and as a warmness at the rounder end. If the gyrations of this life-motor are increased through higher temperatures produced by a brooding hen or the brooding Sun, then in the protoplasm, enclosed by the coarse and fine material of the eggshell, a mysterious weaving and coiling about the yolk begins. A maternalistic expansion and an enveloping, loving and nurturing motion begins, which leads to the consumption of that substance in which has been instilled the irresistible urge so ardently awaited.

It is necessary to read Goethe in order to understand everything that takes place at this moment. This great German poet must also be read with a kinship and affinity for Nature in order to comprehend these mysterious events clearly and to establish the ultimate truth. Otherwise, as Schiller said, one could end up *"pale and insensible at Isis' feet"*, when the vision of truth unveils itself layer by layer. One has also to understand Shakespeare, who described life as a fairytale hitherto experienced and fashioned by fools. In the search for intuitively talented individuals we happen upon the great painters, sculptors and the musically gifted, who were unconditionally faithful to the 'Eternally Transmutable' in Nature and obeyed her inner rhythm. They too were able to initiate impulses.

In this way the flare-up of lightning (enlightenment) also occurred, which lit up everything as broad daylight. With this also came the realisation that this arc of lightning (illumination) ignited and sparked off the impulse to reverse the decline that has arisen from a single misconception.

Index